SCIENCE
GOES TO WAR

BOOKS BY ERNEST VOLKMAN

Legacy of Hate
Warriors of the Night
The Heist
Secret Intelligence
Goombata
Till Murder Do Us Part
Spies
Espionage
Gangbusters

SCIENCE
GOES TO WAR

The Search for the Ultimate Weapon,
from Greek Fire to Star Wars

Ernest Volkman

John Wiley & Sons, Inc.

ISBN: 0-471-41007-1

Printed in the United States of America

10   9   8   7   6   5   4   3   2   1

For Pixie

Clio is the least straightforward of the muses. Her beauty lies in the complexity, not the simplicity, of her truth, which is why her votaries, attentive to the sometimes difficult and winding path they must follow, are sworn to tell stories in order to make the journey easier.

—SIMON SCHAMA

Contents

The Ghost in the Machine

War is the father of all things.

—HERACLITUS, 510 B.C.E.

A t night, when the cold winds sweep through the rugged Judean wilderness overlooking the Dead Sea, it is said that the ghosts of Masada cry out. The anguished wails and lamentations of those who knew they were about to die echo around the treeless, red-brown hills and valleys, a mournful dirge for their fate—and a world that would vanish for two millennia.

In the desolate silence, so it is claimed, the cries of the 960 Zealots confronting their certain doom nearly two thousand years ago in the fortress atop the high plateau sound across the centuries. Perhaps so; the dry desert air has preserved much in this land steeped in blood and history, and it is possible the spirits of the people who died at Masada have been preserved, too. Certainly, the arid climate has preserved much physical evidence, the evidence that so excited Israeli archaeologists forty years ago when they found the fortress of legend did indeed exist—the remains of the high walls that were the fortress's main defense, the deep cisterns that provided an infinite water supply in the event of a siege, the food storehouses to keep the besieged fed for years, the armories where swords, arrows, and spears were manufactured and stored.

There are other archaeological artifacts at Masada, and it is those artifacts that mutely testify to what happened on that terrible day in 73 C.E.—the 100-foot-high dirt ramp Roman besiegers built up a steep cliff to reach the fortress walls 1,800 feet above them; the charred reinforcing timbers for the walls that the Romans set afire with naphtha-tipped arrows; the broken wall smashed by a Roman ram; the worksite below the plateau where the Romans assembled a siege engine; the smooth, round stones scattered everywhere, ammunition fired by the Roman *ballistae* that swept defenders from the walls. And that evidence shows why the Zealots never had a chance.

• • •

The Zealots of Masada were the most fanatical among the Judean revolutionaries seeking to overthrow Roman rule. Six years before, at the outbreak of the Judean Revolt, they had seized Masada, built decades earlier by Herod, from a small Roman garrison and used it as a base to harass Roman occupying forces and murder Jewish collaborators. They probably considered their fortress virtually impregnable, and no wonder: sited atop a plateau whose sides were near-vertical cliffs, Masada represented the state of the art in ancient Middle East fortifications. In addition to its cisterns and food storehouses, which rendered unusable the long-established siege technique of starving out the defenders, the fortress had thick, 20-foot-high stone walls that the Zealots further reinforced with a second set of walls they built 8 feet back, then buttressed with thick timbers. It also had several watchtowers, allowing the Zealots to see any approaching enemy from many miles away. To get near the walls, besiegers would have to make their way laboriously up the cliffs, there to be picked off by defenders atop the walls. And even if the Romans managed to breach the first wall, they would confront the 8-foot-wide gap to the next wall, another killing ground. Given the steep sides of the plateau, there was no way the Romans could bring up a ram to smash through those walls.

Thus the Zealots were justified in feeling they could probably withstand a Roman siege of any dimension, perhaps for years. They gathered their weapons, stockpiled their food, and brought their wives and children into the fortress. Following more than a year of preparation, 960 of them were ready for a Roman attack, which was an inevitability: the Romans were clearly determined to stamp out the revolt, and in the process were laying waste to the entire province, including the destruction of Jerusalem in 70 C.E. By 73 C.E., the Romans had decided it was time to take back Masada, eliminate the Zealot threat, and destroy the Judean uprising once and for all.

The Zealots could not have had any illusions of what was at stake. As all Judea knew, the Romans had destroyed Jerusalem, in the process slaughtering Jews by the thousands and razing Solomon's Temple. So when the Zealots in the watchtowers one summer day saw in the far distance the dust trails raised by the approach of the Tenth Legion, some five thousand men under General Flavius Silva—the very men who had sacked Jerusalem—accompanied by several thousand captured Jewish rebels the Romans had enslaved, there could not have been any doubt: this would be a fight to the death.

The Zealots braced themselves for the kind of assault that had marked war against fortresses and fortified cities for the past eight thousand years in the ancient Middle East: the Romans would doggedly work their way up the steep sides of the plateau, their archers engaging the defenders on the walls while shock troops, carrying scaling ladders, tried to work their way close to the walls. If they succeeded, waves of soldiers would mount those ladders and attempt to overwhelm the defenders on the walls, hoping to find a weak point through which follow-up waves of infantry would pour.

But to what was surely the puzzlement of the Zealots, the Romans initially made no move against Masada. Instead, they simply surrounded the plateau with several camps on the valley floor below and appeared ready only to sit and wait for the defenders to run out of food and water. An odd strategy, since the Romans were aware that Masada had deep cisterns and food storehouses.

Appearances, however, were deceiving, for the Romans were in fact preparing to assault Masada with an ultimate weapon. It was a weapon for which the warriors of Masada, poor farmers and tradesmen—whose knowledge of the military arts did not extend much farther than using their swords, spears, bows and arrows, and stone slings (similar to the one David used to slay Goliath) in hand-to-hand combat—had no defense. And it was a weapon that in the hands of a superbly trained and equipped professional army would mean the Zealots' doom.

The Zealots were about to discover that four hundred years of accumulated scientific wisdom wielded by a civilization scientifically superior to their own could defeat even the most fanatically courageous warriors—a lesson that other civilizations later would learn, at their cost.

• • •

From their posts on the walls, the Zealots watched as 1,800 feet below them, that ultimate weapon was unlimbered in all its dimensions. What was going through their minds as they saw these strange Romans at work can only be imagined. To studiously devout people whose narrow universe did not extend more than a few miles from their ancestral villages and whose grasp of science was limited to basic arithmetic and astronomy (sufficient to tell them when to sow their crops and tally the resulting harvest), the grim Roman soldiers must have appeared as aliens from another universe.

First, they noticed some of the Romans walking around the base of the plateau with strange-looking wooden devices, pacing off distances. Actually, these men were Greek mathematicians on Silva's military staff who were taking careful measurements to determine the best spot to build a ramp up the plateau wall. After working out calculations on their slate boards, the mathematicians told Silva exactly how many hundreds of thousands of cubic yards of dirt would be required to build a 265-foot-long earthen ramp 100 feet high up the face of one cliff to Masada's first wall of defense. They then prepared a schematic showing how the ramp should be built. The thousands of Jewish slaves were put to work, under the sting of cudgels and whips, hauling tons of dirt.

The Zealots also could see several hundred Roman military engineers busily assembling what appeared to be a two-story-high tower on wheels—in fact a large siege machine with a projecting battering ram some 10 feet in length and several feet thick, put together with components the engineers had brought with them. Elsewhere, the Romans were busy assembling several strange contraptions the Zealots had never seen before. However puzzling to the Zealots, the machines, called *ballistae,* were familiar enough to the Romans, who had been using them for two centuries. The *ballista* was a two-armed torsion engine about 8 feet long mounted on a carriage, looking something like a large crossbow. Each arm was tethered to a heavy cord of twisted rope that was joined at the end, forming a sort of large slingshot. The bowstring was pulled back by a winchlike mechanism that also drew back the arms. The machines could fire either steel bolts or round stones; the range was determined by how much tension was used. When the engineers determined that they had winched enough tension to achieve a particular range (which the mathematicians had worked out with precision beforehand and wrote down in the form of tables specifying the ratios between turns of the winch and ranges), the machine was fired by release of a trigger. The stones fired by such machines were round, scientists having determined that such a shape would move through the air with the greatest accuracy and longest range. Indeed, while the engineers were assembling their *ballistae*, Roman artisans nearby were shaping large rocks into perfectly round, grapefruit-size stones for the machines to fire.

With typical Roman industriousness and thoroughness of detail, the full panoply of advanced military science and technology was assembled over the next six months in full view of the Zealots, who must have felt the first chills of fear: how could they possibly defeat such wondrous machines? While below the fortress, the Romans were far out of range of arrows and spears, but the Zealots may have retained a faint hope of destroying or setting afire the siege machine once it came within range of archers and spearmen on the defensive wall. But that possibility quickly evaporated as the Romans finally went on the attack. Foot by foot, as the ramp was built, the siege machine with its battering ram was rolled onto it by several thousand soldiers. Finally, as the machine came within arrow range, the Zealots discovered that the Romans had encased it in sheets of iron, deflecting any missile (and also rendering it fireproof). More terrifying were several *ballistae* the Romans had positioned atop the siege engine, from which stone balls were launched to knock defenders off the walls.

Very soon it was clear to the defenders that their weapons had no hope against such power. Pinned down by the relentless barrage of stone balls whistling through the air, the Zealots could only watch helplessly as the siege engine reached the wall, which crumbled under the blows of the battering ram. Encountering the 8-foot gap, Roman archers inside the siege engine, protected by the engine's iron armor, dipped their arrows in naphtha and set the supporting timbers afire. Then, under covering fire of the *ballistae*, the Romans moved their siege engine forward and smashed through the final defensive wall.

All the while, Silva stood calmly below, issuing occasional orders as the attack unfolded with clockwork precision. In just a few weeks after beginning the actual ascent, at the cost of only a handful of casualties, he had taken the "impregnable" fortress of Masada, in the process wiping out the most dangerous threat to Roman occupation of Judea.

By the time the Romans had breached the second wall of defense, the Zealots undoubtedly realized it was all over. According to the account of the historian Josephus, they decided that their inevitable death (or, more likely, enslavement) at the hands of the Romans was intolerable. With a last, anguished lamentation, they committed mass suicide, drawing lots to determine in what order they would murder each other; their leader was the last, impaling himself on his sword after slaying the other remaining survivor.

• • •

In their final moments, the Zealots prayed fervently to their great god Yahweh for deliverance from the unholy pagans moving irresistibly toward them. But there was no deliverance, and in that moment, the Zealots could only have realized that the world they knew had changed forever: mankind, which for centuries had feared the wrath of God, must now fear the wrath of men.

Ever since the time of Moses, the Zealots and their ancient Hebrew brethren believed, as their Scriptures told them, that the righteous would inevitably triumph; the God of Isaac and Jacob, in his infinite wisdom, would ensure victory of his chosen people over their enemies. Similarly, the Phoenicians believed devotion to their seasonal gods would guarantee victory, the Egyptians believed their gods made them invulnerable, the Carthaginians were convinced that their gods made their armies invincible, the Gauls believed that appeals to their gods would make them victorious in any war, the Druid high priests of Britannia told their people that devotion to the gods of the moon and the stars would defeat any enemy, and the Helvetians believed their gods would make them triumphant.

And yet, all these religiously devout civilizations were destroyed by the implacable military might of Rome. To be sure, imperial Rome had its own pantheon of gods, but no Roman general believed Neptune, or Juno, or Jupiter could be relied upon for success in war. A supremely practical people, the Romans concluded that success in war was guaranteed by superior power, the greatest and most powerful military machine that the world had ever seen. It was a success born of organization; constant training; a professional standing army; technology; and, above all, the utilization of every advance science had made during the previous four centuries.

Other civilizations that had not absorbed this truth paid a terrible price. The half-naked wild Celtic warriors in Britannia, armed only with spears and clubs, believed that adorning their bodies in bright blue paint and making sacrifices to their gods would make them victorious. But the Roman invaders knew better; they knew that no amount of blue warpaint or sacrifices to gods would win a battle—or conquer an entire people. So when the painted Celtic warriors hurled themselves against the Roman legions, they encountered men protected by the latest technical advances in body armor and shields who coolly and methodically mowed down the Celts with volleys of iron bolts fired from *ballistae.*

Like history's other great imperial powers, the Romans understood that as long as one group of men coveted the land, or the riches, or the trade, or the food of another group, war was inevitable. That was a truth established long before the Romans emerged from their tribal huts on a mission to conquer the world.

It all began when Cain slew Abel, in one blow reducing the world's population by a fourth, and ever since, men were slaughtering each other. From the time of the flint ax at the dawn of history, the power to destroy has intoxicated man. The archaeological evidence is sobering: skeletons with embedded projectile points, skulls smashed by clubs, traces of early settlements girded by fortifications and moats to repel constant invasions. From the moment mankind began to walk upright, there was almost constant warfare, a process that accelerated some twelve thousand years ago when the early humans ended their nomadic existence and began to settle in villages to domesticate animals and grow their own food. And as their efficiency in food production grew—exemplified by the invention of irrigation techniques and agricultural surpluses in storehouses—so did covetousness and the will to power. The beginning of war as we now define it grew out of organized theft and the determination of agrarian communities to defend their wealth.

In the centuries since then, war became so endemic that any extended period of peace has been regarded as unusual: in more than

3,420 years of recorded human existence, only 268 have been entirely free of war. It is a statistic that by the seventeenth century would lead the political philosopher Thomas Hobbes to his famous conclusion "The state of nature is a state of war."

Paradoxically, however, this dark thread in the tapestry of human history, while responsible for mankind's greatest suffering, also has been responsible for most of its progress. That is because science has been primarily responsible for that progress. And the greatest spur to science has been war, a relationship that began from the first moment men began to think of better ways to kill each other. The clubs and spears that made man an unrivaled hunter also could be used for war; later, the tools that revolutionized early agriculture were just as efficient as killing tools. The same impetus that led man to improve his life on the planet—a sickle to gather his grain more efficiently, a plow to till more acres of land, a yoke to harness his animals for work, a crude counting system to keep track of his goods— also led him to develop weapons, particularly any that would guarantee superiority (the search for the "ultimate weapon" has a long history). Very early, mankind grasped an essential truth that underlies the history of all warfare: he who has the greatest weapons wins. From the double-edged sword to the thermonuclear missile, success in war has been defined, ultimately, by the ability to develop superior weapons.

And that, in turn, involves greater science. It was the impetus for greater weapons that was chiefly responsible for the birth and growth of science, the uniquely human achievement that has made mankind the master of the planet—and, in the end, threatened to destroy it. Science taught men how to slaughter each other more efficiently, how to blow things up, how to kill men at great ranges, how to subject an entire population to terror, how to wreak total destruction, and how to harness the power of the sun to kill hundreds of thousands of people in one blow. Virtually every modern scientific discipline is rooted in war and the relentless drive for better and bigger weapons to fight it—chemistry arose from the search for more efficient explosives, astronomy from naval warfare's need for efficient navigation, mathematics from weapons ballistics, and metallurgy from the development of edged weapons and guns.

Modern science tends to be uncomfortable with such facts. As a discipline, science prefers presenting itself as essentially neutral, a rational and theoretical inquiry that describes, analyzes, and explains how nature works. In this view, science is dispassionate and rigorously neutral, devoted only to truth. There exist any number of accounts of remarkable minds, totally divorced from the political realities of their time, focusing on the mysteries of nature and arriving at brilliant conclusions that advance man's knowledge in great, single bounds. All schoolchildren are told about Galileo dropping iron weights from the Leaning Tower of Pisa, the apple falling on Newton's head, and the patent office clerk named Albert Einstein who wondered why the glass fronts of shops appeared distorted when he looked at them from a speeding streetcar. In the popular accounts of the history of scientific thought, these seminal events take place in some sort of ethereal realm of human thought, with no connection to the real world in which such minds existed.

It is true to say that there have been great scientific advances born strictly of man's urge to know—Darwin's theory of evolution comes to mind—but what about high explosives? The atomic bomb? Poison gas? Such horrors are categorized in what modern science likes to call "applied science" (also known as technology), the implication being that "pure" science unlocks certain secrets of nature; what mankind decides to do with such secrets belongs in an entirely separate realm beyond the direct responsibility of scientists. The scientific establishment has had the bad habit of trying to have it both ways, taking credit for the applications of science that benefit mankind, while distancing itself from those that more efficiently destroy it. That schizophrenia often has taken the form of a pervasive myth: fundamentally innocent and peaceful science "perverted" by those who use science for destruction.

However, the fact is that the relationship between science and the soldier is a long and intimate one, extending back many centuries. The reason, as physicist Richard P. Feynman once noted, is that science *works;* if a scientist tells a cannoneer to pack an explosive of a certain chemical formula in a precise amount into his gun, elevate the gun to a certain angle, and aim it at a certain point, the resulting explosion will be absolutely guaranteed to hit the target. As Feynman further noted, biological warfare weapons would not exist were it not for the microbiologists who taught soldiers that ordinary germs could be converted into weapons of mass destruction. No soldier ever conceived the idea that gases from an explosion of a mixture of chemicals in a confined space could propel projectiles at great speed and ranges, or that certain wavelengths in the light spectrum could be used to destroy any substance known to man, or that troops in deep fortifications could be destroyed by an admixture of gasoline and chemicals called napalm that literally sucked the air out of their lungs, or that bombarding certain atoms with neutrons would produce an explosion capable of leveling an entire city. Soldiers conceived the requirement for such things, but it was scientists who conceived the solution, at times to their shame.

• • •

"Science" and "scientists" are relatively modern terms. Early societies had medicine men, or wizards, or shamans, or sorcerers (most famously Merlin, the wizard of King Arthur's Camelot), masters of witchcraft and magic whose secret power stemmed from what was presumed to be their access to the will and spirit of the gods. As a result, they had considerable power, which was passed down through generations of successors orally and in great secrecy. Most of that power arose from their mastery of basic science. Babylonian shamans, for example, knew enough astronomy to predict the seasons and devise a crude calendar, immensely powerful pieces of information in an agriculturally based economy, and wizards in the courts of ancient China developed a calendar of the seasons for the emperor. Both developments were kept secret, so that the shamans and the wizards could claim supernatural connections for their knowledge. (The Babylonian shamans were not about to tell the peasant farmers of Babylon that determining the time to sow—reckoned by observing the changing angle of the sun as Earth approached the spring solstice—was a simple matter, and thus talk themselves out of very cushy jobs at court.)

The continual search for more information about the natural world that would further enhance the status of shamans represents the real birth of what we would come to call science. Beginning somewhere around three thousand years ago, as early civilizations developed and were consumed by virtually constant warfare, these first scientists were called upon to use their supernatural powers for an even more urgent purpose: make those civilizations militarily powerful, able to withstand any military threat from without while capable of conquering rival civilizations.

That got early science into the death business, a development that created a very troubling legacy. The blame can be assigned to some very smart barbarians and an even smarter group of Greeks.

I

"The Valor of Men Is Ended!"

Knowledge itself is power.

—FRANCIS BACON

"Come, I will tell you of the ills of the infantryman," the anonymous soldier of the Egyptian Empire dictated to a scribe somewhere around 1400 B.C.E., beginning an extraordinary account of the brutal realities of ancient warfare. It was probably meant as a warning to future generations (including, possibly, his children) about the horrors of serving as a foot soldier, a blunt and angry dose of reality against the antiseptic scenes of military glory the pharaohs liked to have painted on their tombs and temples.

Inscribed on clay texts that would survive to be discovered by archaeologists more than three thousand years later, the soldier's account apparently was written just after he had participated in still another of the interminable military skirmishes between Egypt and the rival Hittite Empire to the east. He tells of being awakened before dawn, "driven like a jackass" by his platoon leader into formation, given a small ration of grain, marching barefoot (with only a little tepid water to drink) for three days over the desert hills to the battlefield, exhorted to the attack ("Forward, O mighty infantryman!"), and, finally, having miraculously survived the carnage, so exhausted by battle he is barely able to walk home.

For the modern reader, there is much that resonates in this unsparing account—the terror of hand-to-hand combat; the anxiety he felt as he saw the Hittite infantry arrayed before him for impending battle; the shock as the two armies, their soldiers screaming war cries, collided in a fury of sword and spear thrusts; the sight of blood and entrails; the screams of the wounded; the officers urging their men onward to greater effort; the generals, safely behind the battlefield and protected in the cocoon of their palace guard, dispassionately ordering men by the thousands to their deaths; and, ultimately, the sheer pointlessness of it all, another bloody battle in the struggle between two empires whose rulers were determined to dominate the ancient world. From the infantryman's perspective, the future was a dark vision of endless other such battles to come, settling nothing. The only certainty was that the price in blood would be paid by the poor infantrymen, the pawns slaughtered in the great game of geopolitical ambition.

The anonymous Egyptian foot soldier was, like his Hittite enemies, almost certainly a poor peasant farmer dragooned into military service whenever the pharaoh or the Hittite emperor decided on a military campaign, usually in the months between the sowing and harvesting of crops. Barely trained, the infantryman was herded with his fellow farmers into mass formations, armed with two light spears and the *kopesh* (a sickle-type sword made of bronze), and protected only by a stiffened linen kilt. In battle, he was sent headlong into the mass formations of his enemy; victory usually depended on how many infantrymen one side or the other managed to hack and stab to death.

The fear felt by so lightly armed a soldier on a battlefield, virtually unprotected against death or maiming in a melee from a spear or sword thrust that might come from any direction, can well be imagined. But the Egyptian infantryman's account conveys the sense of an even greater terror, an ultimate weapon that made him tremble in fear the moment it came anywhere near him. Often, the mere sight of this wonder weapon was sufficient to cause ancient infantrymen to bolt in wild fear and desert their formations, blindly running in a panic that even the threat of execution by their officers failed to stop.

Little wonder why, for the ancient war chariot, among the first great triumphs of human science, represented a military technology that totally dominated the foot soldier and decided the fate of empires for several hundred years. How the war chariot was developed, how it dominated the ancient battlefield, and how it was in turn defeated by even greater science represents a paradigm of how science came to influence war. The chariot's introduction onto the world stage more than two millennia ago began a cycle that continues to this era's Strategic Defense Initiative: an "ultimate weapon," a period of its dominance on the battlefield, the development of a checkmate, and then an endless cycle of bigger and better weapons and counterweapons.

There is another lesson this long cycle has proven: empires and nations that rely too long on the dominance of a particular wonder weapon inevitably become unprepared for the introduction of an even greater wonder weapon unless they have turned science to the task of constantly improving their advantage. The consequences of failing to heed this lesson often have been catastrophic—witness what happened in ancient Egypt.

• • •

In about 1800 B.C.E., somewhere among the barbarian tribes in southern Central Asia, a remarkable scientific achievement took place. The name of its originator (or originators) was never recorded, but whoever it was, the achievement was impressive. The idea apparently arose in studying the early oxcarts, basically crude wooden boxes atop a straight piece of wood connecting two equally crude wooden wheels hauled by a team of oxen. It was sufficient to transport such things as loads of grain, but its limitations as a military vehicle were obvious: too slow, too cumbersome to maneuver, able to travel only in one direction, and offering little protection to any men riding it. The scant archaeological record indicates that there were some attempts to use oxcarts as "war wagons," and the resulting failure seems to have led to attempts to develop a real fighting vehicle.

The final concept behind what came to be known as the war chariot was sheer genius: two hubbed wheels that turned around a fixed axle, the underlying scientific principle of every vehicle on the road today. The wooden chariot, hooked up to a horse instead of an ox, was light, maneuverable, and strong, a technological triumph that included spoked wooden wheels (for lightness) made accurately circular and dynamically balanced to hold up under the strain of rapid motion while carrying several hundred pounds. Additionally, a special harness was devised so that the horse drawing the chariot (or team of horses, for larger chariots that could carry up to three men) would bear part of the weight of the vehicle.

But there was another scientific triumph at the same time that, combined with the war chariot, created the world's first integrated weapons system. That triumph is known as the composite bow. Its unknown developer, after what must have been an extensive and lengthy series of tests and experiments, arrived at a perfect design that combined the shortest possible length with the greatest possible hitting power. It represented a quantum technical leap over the standard bow of the time, a flexible instrument made from sapling wood that was not very accurate, often split, warped in temperature extremes, and had limited range. The standard bow, which had remained virtually unchanged since its invention thousands of years before, suffered several drawbacks as a reliable weapon, but the composite bow changed all that. The key was integrated materials: wood, sinew, and horn glued together for optimum strength and flexibility, able to fire arrows rapidly and accurately at ranges up to 200 yards. The composite bow was shorter than the standard bow, a perfect weapon for the confined space of a chariot.

Following what apparently was a period of field testing and training, armadas of chariots were built, then mated to the skills of an entirely new class of military professionals, the charioteers. Usually the chariots carried two men, a driver and a warrior, both of whom underwent constant training to perfect their skills; the driver was able to move the chariot at high speed when it was time to close for battle while at the same time tightly maneuvering, thanks to the chariot's unique axle design (the ancient equivalent of a small sports car with a supercharged engine). The warrior, armed with a shield, a supply of short spears, a composite bow, and several dozen arrows, would fire his arrows at a rapid rate when he got within 100 yards of a target, then launch his spears as the chariot maneuvered closer. In the event of counterfire, he could use the shield for protection while the driver tried to maneuver out of danger. After a long period of training, warriors could hit targets with near-certain accuracy from a racing chariot.

When the weapons system was finally perfected, the nomads erupted south on a drive for conquest, toward the riches of the lands of Mesopotamia, Egypt, and Palestine. There they encountered peoples totally unprepared for the calamity that was about to befall them.

• • •

Whatever else it may have been, the "cradle of civilization" was a dangerous area in which to live. Since 2800 B.C.E, when the growth of trade created the first empires of early cities woven together by conquest, the ancient Near East was in constant turmoil. In an age of despots, there were endless wars over trade, over economic rivalries, over attempts at aggrandizement, over anything—such as the Sumerian ruler who invaded a Mesopotamian province for the sole purpose of securing enough diorite, a soft stone ideal for carving, so that his artisans could memorialize him with heroic statuary.

No one was safe: despots lived in inaccessible palaces with narrow entrances that admitted only one person at a time; hidden recesses near the entrances concealed men with daggers, ready to stab any hostile visitor. The despots spent most of their time plotting war against other despots or fortifying key strategic points, such as Jericho, astride a critical trade route along the Jordan River, a fortified city of biblical legend girded by 6-foot-thick walls 23 feet high. Palaces and religious temples were decorated with scenes of military horror—victorious soldiers gouging out the eyes of prisoners; piles of corpses; arms and legs hacked off soldiers in battle; defeated soldiers, stripped naked, being led into slavery.

The conduct of these constant wars had remained unchanged for two thousand years: despots consulted shamans on the most propitious time to begin a war. Then ordinary citizens were rounded up, given weapons (which were always stored in government-run arsenals during rare periods of peace to prevent armed popular uprisings), organized into what amounted to an armed mob, and marched off to battle someone else's armed mob. The soldiers, with virtually no training, were equipped only with spears, slingshots, and javelins (short spears). As a result, battle was something of a confusing melee, usually ending when one side got tired of fighting or, more commonly, broke and ran.

For all that period, science had contributed little to the conduct of war, except for the early metallurgists, who solved what appeared to be an insurmountable technical problem. Copper, which was simply lying around in the open, could be fashioned into a metal useful for decorative arts or coinage, but was too soft to be used for weapons. It appears that there was a general assumption that metal weapons would provide a tremendous military advantage over wooden spears, yet the technological hurdle remained: somebody had to figure out a way to transform copper into a substance hard enough to manufacture weapons.

One of history's first arms competitions then developed, a race for a metal useful for weapons. There is no record who finally came up with the solution, but at some point it was discovered that if copper was "alloyed" (meaning "mixed") with other metals, mainly tin, the result was bronze, a strong metal just right for making spearheads, axheads, and swords. Even better, bronze was rustproof and easy to work by a new technical profession the discovery inspired, metalsmithing.

But these new bronze weapons, meant for close-in infantry combat, were of little use when the charioteers came pouring out of the north. Contemporaneous accounts convey some sense of the sheer terror a massed body of unprotected infantry standing in a flat desert felt when they confronted hundreds of war chariots bearing down on them at high speed. The chariots would maneuver around the formations of foot soldiers, striking them down by the dozens from arrows shot by archers at ranges beyond the reach of the infantry's own weapons. The archers could hardly miss hitting at least somebody in such a large and tightly packed target. The foot soldiers' panic would increase as the chariots raced around their rear, cutting them off from their supply lines. At that point, most Near East armies fled in disorder; the charioteers, with superior speed and mobility, ran down and slaughtered individual soldiers.

Within a few decades, the nomads destroyed every military force in the region, and the "people of the chariot" were masters of the entire Near East. Chariots now became the ultimate weapon, and the new kingdoms began a furious military building program to accumulate as many of these wonder weapons as possible. They also sought to build elite units of charioteers. In Egypt, companies of charioteers were stationed at various strategic points in the empire, kept in a high state of readiness and available at a moment's notice to leap into their chariots and head off to block an invasion or to begin one of their own. The first professional standing military organization in history, they were lavishly paid from taxes and booty seized in conquests. Traditional infantry units of conscripts were still used, but now they functioned mainly as adjuncts to the main show, clashes between fleets of chariots. By 1300 B.C.E., when Pharaoh Ramses of Egypt led his army to Kadesh, in what is today northern Lebanon, to battle with the Hittites, he rode at the head of a mighty force of nearly five thousand chariots that fought a huge, swirling battle against thirty-five hundred Hittite chariots. (The battle was a virtual draw, which didn't prevent Ramses from describing it as a magnificent victory in inscriptions he caused to be carved on memorial tablets, a military triumph he modestly attributed to his own brilliant generalship.)

The chariots that clashed at Kadesh represented virtually the same technology that had swept into the Near East several hundred years before. There was no attempt by any kingdom to make improvements, nor was there any effort to devise a new weapon to overcome the chariots' dominance of the battlefield. Certainly, the great Middle Kingdom of Egypt had the scientific wherewithal to do so. Early Egyptian science had produced geometry ("earth measurement"), a remarkable scientific advance that allowed Egypt to build its mighty pyramids and keep accurate accounts of land boundaries made uncertain by the annual flooding of the Nile. And Babylonian science produced the first detailed astronomical observations, enabling Babylon's armies, in a miracle of the age, to find their way at night using star positions as navigation guides.

Why, then, was there no attempt to harness such scientific talent to the development of new weapons? In the absence of records, no definitive answer exists, but the probable reason is that the Egyptians and other kingdoms believed they exclusively controlled an ultimate weapon that had swept all before it. Convinced that this weapon would remain supreme indefinitely, the Egyptians and others became complacent, overlooking a fact of history: at some point, at some time, someone was bound to develop a new, superior wonder weapon.

That inevitability occurred where the chariot kingdoms least expected it, among what the Egyptians contemptuously called "sea peoples," barbarian tribes in Asia Minor and the modern-day Balkans. As in the case of the chariot, no one knows who made the scientific breakthrough, but sometime around 1400 B.C.E., Armenian and Balkan tribes learned that the odd red-brown mineral lying near the surface of the ground could be mined, smelted in a charcoal fire, then quenched and reheated to produce a new miracle metal. It was called iron.

The scientific impetus for this metallurgical breakthrough,

which would come to revolutionize human existence, was strictly military. The barbarian tribes wanted to strike south and conquer the riches of ancient Egypt and Mesopotamia, but so long as the chariot remained the ultimate weapon, their lightly armed infantry stood no chance. True, the barbarians could have built their own armadas of chariots, but that option was unworkable. The chariot kingdoms were rolling in wealth, which afforded them the luxury of simply outbuilding a competitor foolish enough to get into a chariot arms race. Moreover, those kingdoms had large standing forces of highly trained charioteers; many years would have been required to match that quality and quantity—even assuming it could be done.

No, the answer was to redress the imbalance between infantry-men and charioteers; the foot soldier would have to be transformed into a fighting instrument capable of standing up to chariots. Iron solved the problem. Gradually, the metallurgists learned how to transform molten iron into edged weapons and, more importantly, armor. (Later, they realized, possibly by accident, that when charcoal is absorbed in iron during the smelting process, the result was still another wonder metal: steel.) At first, iron was very expensive to produce, but as techniques improved, the metallurgists learned they could produce the metal in large quantities quicker and at a lower cost than bronze, since no alloy was required.

That economic fact turned out to be significant, because it meant entire infantry armies now could be equipped with a whole series of new iron-based innovations at reasonable cost—heavy shields to deflect arrows and other missiles; iron helmets to protect their heads; chain mail to protect their bodies; and long, heavy, double-edged swords capable of cleaving a man in two with one blow.

The military-technological balance swung back to the foot soldier, and when the "sea peoples" began to push southward in about 1200 B.C.E., the chariot kingdoms were surprised to discover that their ultimate weapon was no longer ultimate. Enemy infantry formations stood their ground, parrying chariot-launched arrows with their shields, then attacking chariot warriors with iron-tipped spears and long swords as the charioteers tried to draw closer. Additionally, the charioteers found themselves harassed by another new military innovation, soldiers mounted on horseback (called "cavalry") who attacked the chariots with swords and spears.

The Egyptians and other chariot kingdoms now paid the price for failing to improve their chariot-dominated militaries. The failure extended to the sea: they had done little to develop advanced naval technology to protect their trading ships, an error the "sea peoples" did not make. To the shock of Egypt, which had the largest and most powerful naval force in the region, it was suddenly confronted with a revolution in naval warfare that instantly rendered its own warships obsolete. The revolution came in the form of a new warship that owed its power to several striking scientific advances. Scattered and incomplete evidence suggests that it was the Phoenicians, the Mediterranean's most active sea traders, who probably came up with the idea, an ultimate weapon for the sea. Basically, it was an enlarged version of the standard open-decked warship of the day, powered by a bank of oars. However, the new supership added a large pointed ram at the waterline, capable of staving in the side of any ship then afloat. The key scientific advance that the warship represented lay in its unique brace design for the ram, which kept the ship intact and stable even after the shock of a ramming, much like the shock absorbers on a modern automobile. Additionally, designers of the ship had doubled its speed by a brilliant system of two banks of oars, staggered so that both banks could row at the same time without interfering with each other. And there was also a solid deck over the heads of the rowers, allowing the ship to carry large numbers of soldiers abovedeck who could launch arrows and other missiles against enemy ships or board them.

These military innovations indicated some sophisticated thinking in mathematics, hydrostatics, and metallurgy, the keys to the triumph of the new masters of the Near East. In the process, they had taught a hard lesson about the necessity of using science for practical purposes, a lesson a new empire, Assyria, would put to terrible use.

• • •

The Assyrians, one of the "sea peoples" who had seized the remnants of the Hittite Empire, embarked on an ambitious plan of conquest that involved nothing less than dominating the entire Near East and beyond. To that end, somewhere around 1000 B.C.E. they created the world's first military dictatorship, an entity devoted exclusively to war and conquest. Heeding the lessons of the previous two centuries, they formed an extensive military research and development program that utilized every scientist and technician they could find or lure (with lavish salaries) to the Assyrian capital of Nineveh under injunction to keep Assyria on the cutting edge of military science and technology.

In less than a century, that program created a powerful military machine that made the Assyrians the terror of the Near East. Assyria built a standing army of more than thirty thousand men, unusually large by the standards of the time, since so big a force would require a huge amount of money to feed and maintain. The problem was solved by the aggressive program of Assyrian conquest, with the booty seized from captured lands used to underwrite the cost of the army that had carried out the conquest in the first place. And that army was formidable indeed. The Assyrian military research and development system produced convex shields (the best shape to deflect missile attacks); advanced body armor (including iron breastplates); a new composite bow of greater strength and hitting power; military boots for soldiers; armored chariots; and, in one of the more clever innovations, conical-shaped helmets (which deflected arrows that had been fired at high trajectory). Combined with a relentless program of constant training, this technologically superior military machine in the hands of a succession of Assyrian military dictators ruthlessly swept all before it.

The greatest triumph of Assyrian science was the *helepolis*, a wonder weapon that terrorized the Near East for several hundred years. The first known siege engine, it was a 100-foot-high siege tower mounted on four huge rollers that made it virtually impossible to tip over. It required hundreds of men to move this mighty machine just a few feet, but once in position before a besieged city, it provided an immense advantage to the Assyrian attackers. The entire machine was wrapped in dampened animal hides to checkmate fire arrows launched by the defenders, while inside, Assyrian archers picked off defenders by firing arrows through slots. The thick, fireproof top of the machine was hinged; when the Assyrian soldiers contained inside were ready to carry out their final assault against the walls, the top was opened onto a wall to serve as a ramp.

An Assyrian siege was a masterpiece of coordinated arms and run with mathematical precision, often aided by another innovation of Assyrian science, the ancestor of the modern military tank. This machine, designed to smash through the thick gates of walled cities, was mounted on six large wheels, with a turret for observation. The business end was an adjustable battering ram that could be aimed at just the right angle to achieve penetration, either horizontally or vertically. Protected by plates of leather and wicker, the machine was wheeled into position and then, guided by adjustments called out by the man inside the observation turret, was pushed against the targeted gate or wall.

Once inside a besieged city, the Assyrians showed no mercy; like all succeeding military dictatorships, they believed in the use of terror as a weapon. The entire population of a captured city was slain or mutilated; word of their fate was spread far and wide by Assyrian messengers as a warning to those thinking of opposing the Assyrian military machine. Anyone doubting Assyrian mercilessness needed only to read the boastful and bloodcurdling inscription the Assyrian military dictator of 900 B.C.E., Ashurnasir-pal, chiseled in stone following one of his victories: "I cut off their heads. I burned them with fire, a pile of living men and heads over the city gate I set up. Men I impaled on stakes. The city I destroyed, devastated. I turned into mounds and ruin heaps, the young men and maidens in the fire I burned."

Such ruthless imperialism eventually led to Assyrian domination of Egypt, Syria, Mesopotamia, and Palestine, but it carried a fatal flaw: the system worked only if the dictator had the requisite leadership qualities of supreme power that enabled him to totally dominate a huge military machine, a sprawling empire, and a state apparatus directed solely toward conquest and the continuing improvement of its military arm. But dictators with that kind of ability were rare, and by the seventh century B.C.E., the Assyrians were beginning to run out of them. Under weak leadership, the Assyrian Empire started to fray, and by 610 B.C.E., it collapsed altogether.

Another reason for the Assyrian collapse was the empire's inability to maintain its military edge. The Assyrians were convinced that draconian security measures—such as decreeing that their metallurgists must never leave Assyria and telling scientists they were forbidden to work for any other empire under pain of death—would keep the latest advances in military science in their hands. But that goal proved impossible; unless the Assyrians were prepared to lock up their scientists and technicians in a dungeon, there simply was no way to keep scientific and technical talent from slipping away to the lure of even more gold offered for their talents by Assyria's enemies to match Assyrian military power. The Phoenicians, wealthy from the profits of their trading empire, were willing to pay top dollar for the kind of talent that could help maintain their dominance of Mediterranean trade, and the Carthaginians, a rising military power in North Africa, also were willing to lavish money for scientific help in improving their army (for one thing, they were looking for scientists they hoped knew enough about elephants to transform the animals into military assault weapons).

Given this avid market, an explosion of applied science was about to take place. When it detonated, however, the explosion occurred not in the Near East, but farther west, in a patchwork of quarreling city-states known as ancient Greece. And what happened there would change the world forever.

• • •

Somewhere around 600 B.C.E., just as the Assyrian Empire was collapsing, a group of men in the Ionian area of Greece who called themselves "philosophers" came to a revolutionary conclusion: the world was not, as their culture had taught them, the playground for the Greek gods, in which man's fate was decided by the whim of one god or another. Man, they concluded, had been born with a rational mind, an instrument that should be used to apply reason as a means of comprehending the "ways of nature."

Modern science was born at that moment, for the men history later called pre-Socratic philosophers immediately set about attempting to divine the mysteries of how the world and the universe worked. They and their successors—Plato, Aristotle, Socrates, and other giants of rational thought—would make great leaps forward in mathematics, philosophy, and logic, the fundamental auxiliaries in the construction of empirical science. Their range of scientific accomplishment was breathtaking—the fundamental concepts of astronomy, mathematics, mechanics, hydrostatics, physics, physiology, and the causes of motion of material objects.

The Greek scientific accomplishments are often categorized as "pure science," the result of human minds, freed from the shackles of religious dogmatism, attempting to understand how the world worked, for the pure joy such comprehension offered. There was an element of that, but the major spur to the explosion of Greek science was the environment in which it was spawned. Ancient Greece was a polyglot of city-states that were at constant war with each other and hostile neighbors. War was so endemic that Greece's greatest political thinkers, Plato and Aristotle, took the persistence of war for granted, the inevitable result of what they regarded as mankind's acquisitive and aggressive nature. All wars, they concluded, were essentially varieties of man's drive for power, and given that the urge for power was an ingrained part of human nature, war was as natural as breathing. Greece's greatest historian, Thucydides, added a geopolitical explanation: given the proximity of competing empires, city-states, and trading powers in the Mediterranean, constant war was inevitable because everybody wanted what everybody else had.

The Ionian philosophers, the first great Greek scientists, shared this view. More importantly, they were loyal Ionians. So when Thales of Miletus, Greece's first great scientific thinker, was asked by his native Ionian city-state to apply his great mind to his birthplace's desire to become a naval power, he obliged by using his mastery of geometry to provide Ionian naval captains with a system that allowed them to calculate precisely the distance from ship to coast. It was a critical advantage in a time when ship captains, lacking adequate navigation tools, hugged the shore, making their way by landmarks (an often dangerous system when bad weather obscured those landmarks and caused ships to crash into unseen rocks or other navigational hazards). He trumped that with a critical insight from his astronomical observations: Earth appeared to revolve on an axis that pointed to a "fixed star" in the heavens, the only one that did not move in the night sky. And since that star—Polaris, in the Ursa Minor constellation—was a constant star in the north, sailors could now orient their direction on the sea by reference to their positions in relation to Polaris (an insight that created the science of navigation). Several of Thales's colleagues made even greater strides, among them devising a system of spherical coordinates that would allow the navigator of a ship to locate with precision any point in the sky, from which he could then determine how far he was from the nearest land.

Similarly, when Athens was threatened by a Persian invasion in the fifth century B.C.E., scientists loyal to Athens went to work in a large-scale scientific and technological effort to give the city-state an advantage in the struggle against a much more powerful opponent who had a vastly larger army. That effort resulted in a nasty technological surprise in 490 B.C.E. for the invading Persian army when it encountered the Athenian army drawn up on a plain near Marathon. The unarmored Persian infantry, armed with spears and swords, was intended primarily as a mop-up force after Persian archers opened a battle by firing clouds of arrows against enemy infantry formations. But to the shock of the Persians, the Greeks at Marathon were impervious to the arrow barrage, thanks to new bronze armor. Athenian scientists had carefully studied the standard Persian infantry tactics and concluded that the key was the Persian archers; if their deadly barrage could be blocked somehow, then the Greek infantry, superior in hand-to-hand combat, would decimate the Persian infantry. Working with bronzesmiths, the Greeks developed a new body armor system that featured a helmet that protected an infantryman's nose, cheeks, and neck—the most common targets of slashing swords—along with bronze greaves to protect legs, and new lightweight, but strong, shields. When the Persian archers opened fire, the Greek infantrymen formed a "turtle" of shields that deflected the arrow barrage, then attacked the unarmored Persian infantry, slaughtering six thousand of them.

Both the Ionian and Athenian triumphs stemmed from a military research and development effort utilizing the best of Greek science. The concept had been borrowed from the Assyrians, and was part of a general pattern of Greek thought: the Greeks borrowed freely from wherever they found anything of scientific value—the Egyptians, Babylonians, Carthaginians, and Phoenicians. The Greeks, however, made two important adaptations. One, instead of restricting science to a priesthood that closely guarded its secrets—as in Babylon and Egypt—the Greeks freely circulated their scientific insights, believing that sharing information among scientists was the best way to stir further development, the reason why virtually all early science texts were written by Greeks. Two, the Greeks believed that although "pure" science was the ideal, science nevertheless had a responsibility to the state; in the event of war, scientists should use their abilities to help preserve the state. In this connection, the Greeks drew a distinction between theoretical science and utilitarian science—or, in more modern terms, applied science.

It was applied science that did the most to spur the development of Greek science in general, especially mathematics. And that was because of events that had occurred many centuries before, at a city called Troy.

• • •

All Greeks knew about the siege of Troy in 1184 B.C.E. by the Achaean Greeks, thanks to the Greek national epic, Homer's *Iliad.* However inspiring this heroic saga was to generations of Greek warriors, the more hardheaded rulers of Greek city-states saw it in more practical terms. The siege of Troy had taken eleven years and, not including the cost of the giant gift horse Homer said the besiegers used to finally take the city, the drain on the Achaean treasury was ruinous, to say nothing of the terrible casualties. Such hollow victories represented the wrong way to run a war. And considering the fact that Greek city-states were all arming for the next war and simultaneously building up their defensive fortifications, wars and sieges were inevitable. The question, then, was how to conduct those wars without getting bogged down in the kind of protracted struggle that had, in the end, ruined the Achaeans.

The problem was that Greek warriors who attacked Troy confronted protective walls several feet thick and 22 feet high, against which they had only their swords, spears, and scaling ladders. The solution, clearly, was some kind of wonder weapon that would enable besiegers to capture a city relatively quickly, without having to spend years and a lot of money and blood. Ideally, such a weapon would fire missiles at long range to knock defenders off the walls or keep them pinned down while assault troops would be able to breach the walls. (And, on the other side of the coin, the same kind of wonder weapon would enable defenders to beat off an attack by even vastly superior numbers by preventing them from approaching the defensive walls.)

The answer seems to have arrived simultaneously in a number of Greek city-states: torsion artillery, the most wondrous and frightening weapon the world had ever seen. The idea apparently was borrowed from the composite bow, which the Greeks enlarged into a machine the Romans later called a *ballista.* As noted earlier, the *ballista* was essentially a large bow (or, perhaps more accurately, slingshot) that could shoot iron darts or grapefruit-size stones at

high velocity. Also developed was the *onager,* a *ballista*like machine that had a firing arm under tension instead of a rope bowstring; it was used to fire large rocks.

The Greek scientists who invented what they called "engines"
(from which we get the modern word "engineer") demonstrated an
astonishing range of scientific accomplishment in devising them.
Even today, the sight of these machines inspires awe at the achievement. In a time long before computers or precision instruments, the
Greeks worked out the formulas that measured the energy stored in
ropes twisted into skeins and converted that into various ranges.
They also developed mathematical firing tables for the engineers
who operated the pin and ratchet mechanism that twisted the ropes
into the proper tension for firing; by merely consulting the tables,
engineers knew the precise number of turns to make on the ratchet
to achieve any particular range. Even more ingenious was a system
of large perforated washers set at predetermined points in the rope
skeins to regulate the torsion (and thus the range), precursor of all
modern artillery guidance systems.

Given all the work needed to develop such technology, it is little
wonder that science in Greece, especially mathematics, advanced
farther than in any other civilization. Designing such machines required development of advanced concepts in physics and mathematics that could work out the optimal designs for maximizing the
kinetic energy of the projectiles fired by the machine in conjunction
with the energy of the torsion springs. It also required an advanced
grasp of the sciences of ballistics and elasticity.

All these advances were summarized in the ultimate Greek torsion artillery, the *katapoltos,* a machine that would remain in use
until the invention of gunpowder many hundreds of years later. The
katapoltos (or "catapult," to use the later term) was a giant-size version of the *onager.* A crew, using a double set of large ratchet and
pinion mechanisms, was needed to winch the firing arm into position, where it was armed with stones weighing up to several hundred pounds. Fired by a trigger, the largest versions of the machine
could launch heavy rocks at targets up to 500 yards away. Adjustments to the firing mechanism would allow a flat trajectory, useful
for knocking down walls, or a high trajectory, which would launch
stones over the walls into a besieged city. (Alternatively, the cata-

pults could launch an early version of "Greek fire," a combustible mix of sulfur and other chemicals that, when set afire and launched into a city, exploded into flames on contact, the ancestor of napalm.)

The Greek scientists also were enlisted to improve siege towers, and by the beginning of the fourth century B.C.E. they had developed a 9-story-high monster. This massive machine had firing ports protected by an internally guided system of shutters, through which archers fired iron bolts from still another development of Greek science, a crossbow, to attack defenders on the walls. To prevent the siege engine from being set afire, it included a system of water tanks from which soldiers inside could draw water to douse fires. The entire structure rested on eight huge iron-tired wheels and required the brawn of thirty-four hundred soldiers to move. It was usually accompanied by an entirely new siege machine that the Greeks nicknamed "the tortoise"—a huge, covered, wheeled shed that carried a thousand soldiers and had battering rams 180 feet long.

"Oh, Heracles," the Spartan king Heracles wailed aloud as he watched these mighty siege engines slowly advancing on his city, "the valor of men is ended!" Not entirely, although it was clear that science now had begun to rule the battlefield. As the Greek city-states continued to war with each other and with foreign enemies, the machines of war became bigger, more ingenious, more deadly, and more terrifying. The scientific-technical race for bigger and better machines reached a pinnacle of sorts in 305 B.C.E., when the island fortress of Rhodes came under siege by a 10-story-high Greek siege machine. Covered by catapult fire and protected by troops in tortoises, the Greeks assumed that their siege machine, slowly inching its way toward the fortress walls, would ultimately prevail against the much smaller machines and thin numbers of the defenders. But the rulers of Rhodes had their own secret weapon: science. Some time before, anticipating the Greek siege, Rhodes hired (for a whopping fee) Kallias of Arados, the most brilliant among the Phoenician scientists. Kallias had studied the Greek siege machines and devised a defense that turned out to be, prosaically enough, a hole in the ground. He was among the leading thinkers on the new science of mining, and that turned out be the key to defeating the Greek siege: Kallias carefully calculated exactly where the siege engine would be moved, then put hundreds of miners to work digging a deep tunnel that extended some distance from the fortress. Just at the exact spot he predicted, the huge siege machine collapsed the roof of the tunnel and sank into a deep hole, immobile and useless. The Greeks immediately abandoned their machine, gave up the siege, and moved on to an assault against Thebes, where they built an even bigger machine, so large it required four thousand men and two months merely to move the monstrosity a quarter of a mile. (Meanwhile, the ever-practical people of Rhodes dismantled the huge Greek siege engine left behind; sold off its timber, bronze, and iron; and with the money built the Colossus of Rhodes, one of the ancient world's Seven Wonders. It was intended as a tribute to their sun god, whose intervention was credited as saving the city—much to the annoyance of Kallias, who knew better.)

All this internecine warfare among the Greek city-states finally ended during the midfourth century B.C.E., when Philip of Macedon conquered them and created a unified Greek Empire. But his ambitions went much farther: destruction of Greece's greatest rival, the Persian Empire to the east. Philip was assassinated before he could realize this ambition, but his son Alexander would achieve it—and a lot more, besides: the conquest of virtually all the civilized world.

In 343 B.C.E., when Alexander was thirteen years old, his father assigned him a tutor in science, the one field of human endeavor Philip believed was the guarantor of Greece's future greatness. The man who got the job, Aristotle, considered the greatest mind of the ancient world, would fill the young king's mind with the accumulated wisdom of Greek science. As a result, Alexander became history's first scientifically literate conqueror, and that fact accounts for much of his success. He embarked on a remarkable campaign of conquest that not only destroyed the Persian Empire but also went on to conquer Egypt, Mesopotamia, India, and Central Asia. All fell before a military machine without parallel, the most powerful and efficient army the world had ever seen to that point. Its cutting edge was Greek science.

Alexander institutionalized science, gathering scientists in state-run research organizations, arsenals, and shipyards, and underwrote an extensive military research and development effort directed toward providing him with advanced tools of war. In the hands of a man like Alexander, an authentic military genius, these fruits of Greek science overwhelmed every military force that tried to oppose him—powerful shipboard battering rams for his ships; marine

boarding bridges for amphibious assaults against island fortresses;
advanced torsion catapults that could fire huge rocks at ranges of
800 yards; *parabolos,* a new rapid-fire stone-throwing assault machine; small catapults that operated on compressed air; and the
most wondrous of all, an automatic-firing *ballista* with a mechanism that fired arrows sequentially without the necessity of rewind-
ing the torsion ropes after each shot, the ancestor of the machine
gun.

Alexander died prematurely of disease in 323 B.C.E., but before
his death, he made what would turn out to be the greatest contribution to science in history—the *museion* at Alexandria, the Egyptian city named in his honor. Alexander's idea was to create the
penultimate scientific research institute that would join Western
and Eastern science in an effort to solve all practical problems of
running the new Greek Empire and ensuring that it remained
supreme over all possible competitors. Its mandate included engineering, navigation, astronomy, geography, road-building, determining land boundaries—and the machines of war. He chose a site
on Egypt's northern coast, near the Nile River, for the institute, a
junction of the trade routes of Asia, Africa, and Europe, and therefore the virtual center of the civilized world. Its construction underwritten by riches seized from the Persian Empire, a magnificent
seaside trading city was built, centered around the *museion.* There,
supported by lavish funds, the cream of Greek science was gathered
to conduct research, to expand knowledge in one of the hundred
classrooms for classes taught by men called "professors," and to
work out scientific problems in one of the many laboratories. The
crown jewel of the *museion* was its library, where more than seven
hundred thousand manuscripts were eventually stored. By Alexandrian decree, all ships entering the harbor at Alexandria were required to turn over any scientific works on board; these were copied
by platoons of scribes, creating what still ranks as the greatest scientific library ever. All living expenses of the scientists working at the
museion were underwritten by the state. They learned that they
could hardly think of a line of research that would not be funded; if
it had anything to do with benefiting the state, there was a certain
guarantee that the state would throw money at it.

Given such largesse, the *museion* created a flowering of science that was not to be rivaled until the Renaissance, more than eighteen hundred years later. It was where Heron made important advances in the study of gases (he was the first to realize that steam in a closed cylinder could move machines); Eratosthenes calculated the circumference of the Earth (he turned out to be off by only a few hundred miles); Hero built a primitive reaction turbine; Ptolemy developed his theory of how the solar system worked (although he mistakenly put the Earth at the center of it); Hipparchus devised the idea of latitude and longitude; and Philon wrote his seminal treatise, *Elements of Mechanics*, the basis for all subsequent mechanical sciences.

But it is important to understand that all these great scientific conceptual breakthroughs grew out of the main reason why the *museion* existed in the first place: to advance the power of the state. Hipparchus and Aristarchos made important advances in measuring land surfaces, the basis for surveying and cartography, but these advances were the "pure science" results of their work to develop better navigation systems for warships and efficient means to determine the empire's boundaries (so the state would know the best sites to station garrison military forces). Ptolemy's discoveries about the motions of heavenly bodies were the result of his "star tables," devised to provide navigators of warships with the ability to steer at night, using star positions; Philon's insights into mechanics stemmed from his work on developing more efficient catapults. And one of the greatest mathematical discoveries made by Greek mathematicians at the *museion*—the extraction of cube roots—was the result of work to devise "catapult formulas," the optimal dimensions for any ballistic machine or projectile.

The simple fact is that as long as the state paid the bills, the scientists of Alexandria were compelled to do the bidding of the state, the basis for what would become known as the Golden Rule of all future scientific research: he who pays the gold makes the rules. This tension between pure and applied science troubled some of the Greeks, who had real trouble trying to resolve it. The problem, fundamentally, was that the men of science were also, in Plato's famous phrase, political animals. Aristotle insisted that science should be devoted strictly to the accumulation of "pure" knowledge, but he also was a fervent Greek patriot whose hatred for Greece's Persian enemies bordered on the rabid. He passed on that visceral hatred to

Alexander, making it clear that his young pupil's mission in life was to rid Greece of the Persian threat. Presumably, Aristotle had no objection if Alexander used science to achieve that goal. Such great Athenian minds as Socrates and Plato argued that science should be divorced from any social or economic objectives, but never explained how the utility of science was to be put aside when the Persian armies showed up, determined to wipe Athens off the map.

Judging by the historical record, a number of Greek scientists were uncomfortable working on engines of war, which they regarded as a perversion of the true philosophical goals of science. It was a discomfort best illustrated by the life of the man often regarded as the greatest Greek scientist of them all, Archimedes of Syracuse.

• • •

Archimedes had studied in the mid-third century B.C.E. at Alexandria, where his studies of mechanics first resulted in what became known as the Archimedean screw, a helical pump to move water from the Nile into irrigation ditches; the scientific principles behind it rest today in every drill bit and screw. From there, Archimedes moved on to a number of other epochal discoveries in mathematics and mechanics, most famously his insight that the density of an object in water can be determined by comparing its weight to the amount of water it displaces, the foundation of all hydrostatics and naval science. (Among the many legends about Archimedes' life, it is said that after making this discovery while lowering himself into a bathtub, he jumped out of the tub and ran through the streets crying, *"Eureka!,"* the Greek phrase meaning, roughly, "I have found it!")

Archimedes became equally famous for a series of incredible engines that used pulleys, levers, and fulcrums to move great weights. According to the Roman historian Plutarch, Archimedes considered such work "ignoble and vulgar," and there is no mention of it in the fifty extant scientific works he wrote. He instead wrote about his fundamental discoveries in mathematics, chiefly formulas on finding the areas of various geometric figures and determining the volumes of spheres. But however disdainful Archimedes might have been about the practical uses of his scientific discoveries, he was a fervent Syracusan patriot. So when Hieron II, the ruler of Syracuse, begged his help in 215 B.C.E. at the moment of the city's greatest crisis, Archimedes put his scientific genius to work in the service of war.

Syracuse, a Greek colony in modern-day Sicily that occupied a key strategic position athwart Mediterranean trade routes, had made the error of supporting the Carthaginians in their war against Rome. The Carthaginians were defeated, and now the Romans had come after Carthage's ally Syracuse. A Roman invasion fleet of eighty ships showed up in Syracuse's harbor to begin a blockade while some fifty thousand Roman troops prepared to besiege the city.

Appointed general of ordnance for the city, Archimedes went to work. He designed a number of advanced war machines, including a huge swinging crane that hurled 600-pound leaden balls; rapid-firing catapults that shot bundles of Greek fire; and, if some accounts are to be believed, a system of giant mirrors that reflected concentrated sunlight to burn ships. For three years the Roman besiegers threw themselves at this array of military technology, to no avail: Roman ships were smashed to pieces and Roman troops were cut down at long range by high-velocity fire from catapults Archimedes positioned atop the city's defensive walls.

Finally, in 212 B.C.E., while the Syracusans were celebrating a religious festival, the Romans discovered an unguarded gate, and the city fell. Roman soldiers who poured through the gate found a half-naked elderly man sitting in a bed of sand, absorbed in drawing geometrical shapes. When one of the soldiers stepped onto the sand, the old man snapped at him, "Keep off, you!" Enraged, the soldier immediately ran his sword through Archimedes of Syracuse, then joined his comrades in an orgy of looting and killing that destroyed the city.

It is doubtful that the Roman soldier recognized the famed Archimedes, or would have spared his life even if told the identity of the scientific genius he was about to slay. By the time of Archimedes' death, Rome had emerged as the great imperial power following the gradual breakup of the Alexandrian Empire, frittered away by Alexander's quarrelsome successors. But unlike its historical predecessors, Rome did not depend on science to gain mastery of the known world. In fact, the Romans virtually ignored science, which would enter a great slumber until reawakened many centuries later.

Ever since they had emerged in 600 B.C.E. from their peasant community in Italy and went on eventually to extend their dominion over an immense empire extending from the Irish Sea to the Persian Gulf, the Romans disparaged science as an idle plaything of indolent philosophers. They were a hardheaded, practical people for whom only utility counted. The great Roman military machine used the same war machines and innovations invented by the Greeks, but never sought to improve them. Busy building an empire, the Romans sought only wealth and power, and had no use for anything that did not contribute to those goals. For that reason, the Romans were great engineers, constructing 53,000 miles of roads, a network of aqueducts, and great buildings (inventing concrete in the process). The real cutting edge of the Roman legions, aside from their superb discipline and constant training, was engineering. The greatest Roman military leader, Julius Caesar, utilized brigades of engineers who provided the edge for some of his biggest triumphs—such as his siege of Alesia in Gaul in 52 B.C.E., when his engineers excavated 2 million cubic yards of earth in two weeks. Later, when the Helvetians considered themselves safe from Caesar's army behind the barrier of the Rhine, they were stunned when Caesar's troops fell upon them after crossing the river on a bridge his engineers built in only twenty-four hours.

In a slave-based economy, the Romans expected others to do all the heavy lifting for them, including science. Greek scientists and technicians enslaved when Rome conquered ancient Greece took care of maintaining such war engines as the *ballistae,* and to serve as scientific advisers to Roman armies. The Romans had seized hundreds of thousands of scientific texts in their conquest of Greece and Alexandria but made no effort to translate them into Latin. Many of them wound up in the villas of prominent Roman officials and merchants, there to serve as decorations attesting to the owner's status as a man of learning, although he had not read a word of them. (One Roman patrician had sixty thousand Greek manuscripts on display in his living room, but since he also had twenty-two concubines in residence, there is strong doubt that he ever found the time to read.)

As Alfred North Whitehead later noted, "No Roman ever lost his life because he was absorbed in the contemplation of a mathematical diagram." Just so; the Romans had no interest in science, military or otherwise, because there was no reason for such interest. There were no major military threats to spur scientific development, no necessity to examine the laws of nature (the Romans believed they already understood them), and no reason to delve into how the world worked (how would that build bigger villas, provide more glory, and increase wealth?). The Romans were so convinced of their approach to life that a prominent Roman official, Frontinus, in 100 C.E. would write that there was no necessity to invent any further "engines of war," since Roman arms were already more than sufficient to rule the world. And when the Romans encountered a rare situation where they came up against superior technology, they simply put their engineers to work adapting it. For example, in 260 B.C.E., confronted by superior Carthaginian warships that used the advanced science provided by the wizards in Alexandria, the Romans reacted characteristically. Instead of developing their own naval science, the Romans captured a Carthaginian warship, then in a feat of what today would be called reverse engineering, took it apart piece by piece, figured out how the ship was built, then constructed an exact copy. In three months the Romans built a fleet of 220 warships exactly like it and swept Carthage from the seas.

By 250 C.E. the Romans looked upon a world that was completely dominated by Rome. The only remote threat were the barbarians at the outer borders of the empire. They were fierce warriors but primitive ones, without the siege engines, the catapults, the *ballistae*, and the phalanxes of highly trained Roman legions armed with their strong shields, body armor, javelins, and deadly *gladus* (a superbly crafted short, double-edged steel sword). But if the Franks, the Angles, the Vandals, the Teutons, and the Goths had no such technology, the fruit of thousands of years of scientific development, they nevertheless had brains. And those brains began to think of some kind of technological edge that would equal the odds between themselves and the Roman army.

While the complacent Romans rested content behind their walls and forts guarding the empire's outer edges, the barbarian tribes developed a technological breakthrough. The inspiration for it came from the Scythians of Eastern Europe, a nomadic tribe that virtually lived on horses. The Scythians, intent on prolonging a horse's useful life, sought to make the animal's burden of carrying a man easier. They invented a crude saddle, consisting of two cushions, stuffed with horsehair, and connected to the horse's body by straps. That meant a rider's weight now rested over the horse's dorsal muscles instead of chafing the animal's spine. The Scythians added cloth loops to the straps so the rider could hook his big toes into them, making control of the horse easier and allowing for easier mountings with less impact on the horse's spine.

The barbarians refined this system into military technology, in the process revolutionizing warfare. They added a harness to the Scythian system and replaced the cloth loops with iron stirrups, which now meant that a rider could hold himself steady atop a horse while either shooting arrows or wielding a sword. There were further refinements and constant training to perfect the tactics of the new heavy cavalry, and by 250 C.E., the barbarians were ready: they began to push into the borders of the Roman Empire. A series of border wars broke out in which the barbarians continued to perfect their new military arm, most notably adding a long wooden lance to the cavalry's armory.

In 378 C.E., the crisis arrived. Tired of the barbarian incursions, the Romans decided on one great battle that would break their enemies forever. The chosen target was the Goths, who had overrun the Danubian provinces, crossed the Danube, and invaded the Balkan peninsula. The Roman emperor Valens personally led a huge Roman army of nearly fifty thousand men to take on the Goths, who had gathered their own forces for a showdown battle. It came near Adrianople, when the Romans, sighting the main Goth encampment, went on the attack. Suddenly the Romans were assaulted by great swarms of Goths on horseback who tore into Roman formations from all sides. Crashing into the legions at full gallop, Goth lances hit the foot soldiers with such force, the weapon skewered three or four men in one thrust. Other horsemen fired volleys of arrows or cut and slashed with swords. Roman soldiers, pressed into a dense mass by the horsemen who surrounded them, couldn't unlimber their swords or spears. In less than three hours, the Goths killed more than forty thousand of them, including Valens himself.

Adrianople marked the end of Roman military dominance; within a few decades, Rome's power would be broken completely by the new ultimate military weapon, the warrior on horseback, enabling the leader of the Visigoths to make good on his vow to turn Rome into a sheep pen. The new masters of the civilized world were the barbarian tribes who now poured into Europe. They would plunge the continent into the Dark Ages, very nearly extinguish the flickering light of science, and come within a hairbreadth of collapsing into the graveyard of history that contained the forgotten empires of Sumer, Babylon, Egypt, and Assyria.

But everything changed when some of them one day many years later entered a city in Spain and felt a distinct whiff of fresh air.

2

Bride of Faith

*A prince ought to have no other aim in thought
nor select anything else for his study than war and
its rules and discipline; for this is the sole art that
belongs to him who rules, and it is such force that
it not only upholds those who are born princes,
but often enables men to rise from a private
station to that rank.*

—MACHIAVELLI, *The Prince*

It had rained heavily the night before, so by dawn of October 25, 1415, the field near the small village of Agincourt in northern France, where the English army had camped, was a soggy mess. The small, ragged force of five thousand men and their king, Henry V, woke to a light mist and chilly air, worsening their misery. Most were sick with dysentery and bronchitis, the diseases that had ravaged Henry's army since it had landed in Normandy six months before in still another phase of the long struggle between England and France that would become known as the Hundred Years War. Of the twelve thousand men in the original English expeditionary force, fewer than half remained—and most of them were sick. The rest were exhausted by months of inconclusive maneuvering and countermaneuvering across the French countryside, marching 14 miles a day in a vain attempt to force a final, showdown battle with the French in a war that seemed to have no end. Critically short of food, they were forced to pillage, angering French peasants, who were exacting revenge by murdering stragglers and keeping a pursuing French army well informed of English strength and direction of march. So well informed, in fact, that the English that morning discovered that the French army had found them. Now there would be that showdown battle, but the odds were three to one in favor of the French.

As the morning mist began to clear, the English saw that a French army of fifteen thousand had taken up positions directly across the English route of march. While the French watched contemptuously from a small rise, Henry deployed his troops for battle. Judged strictly in terms of numbers of troops and weapons, the English situation was hopeless. Against Henry's small force was arrayed the very cream of European military might and the cutting edge of contemporary military technology. At the front of the French dispositions were footmen, infantry soldiers armed with swords, battle-axes, pikes, and maces (spiked iron balls swung from the end of a chain). And behind them, arrayed in battle order waiting for the order to sweep down and destroy everything before them, was the real heart of French military power.

In the morning sun, they were a magnificent sight—five thousand knights, the glittering stars of France's aristocracy in their polished armor. Fluttering in the breeze were their magnificent flags of coats of arms, signifying the various French noble dynasties whose honor they had sworn to uphold, along with the brightly colored plumes atop their face armor. They waited impatiently for battle atop magnificent horses, specially bred for the size and strength necessary to carry the weight of man and armor. These men and horses represented a weapons system that since the time of Adrianople some thousand years before had totally dominated warfare, a frightening combination of power and speed that could destroy any body of infantry in minutes.

Certainly, the English that day appeared to present very little threat. Looking like the shabbily dressed peasants they were, the English had no knights, no war machines, no plumes or bright silk flags. All but eight hundred of them were archers, but as the knights well knew, arrows presented very little threat to an armored knight. The French knights wore the latest triumphs of German armorers, overlapping metal plates covering chain mail, the kind of protection that would stop an arrow fired even at close range. The French war machine was the best in Europe; those archers who had now taken up position in ranks at the head of the English deployment, spiking dozens of arrows in the ground at their feet for a ready ammunition supply, were doomed. The infantry would push them back and then, at the time of the greatest disorganization, the knights would sweep in and finish them off.

Scornful of their motley English enemies, the French infantry opened the battle by carrying out the ultimate gesture of contempt in fifteenth-century warfare: they advanced to within 200 yards of the English troops, halted, turned their backs, dropped their pants, and bent over, offering their backsides.

At that moment the French should have realized that Agincourt would be no ordinary fifteenth-century battle, for the English archers unlimbered a distinctly odd-looking weapon these French troops had not seen before, a bow nearly a foot taller than the man using it. And although the French footmen assumed they were out of range, within a minute hundreds of them were writhing in agony from arrows shot into their bowels. The others turned to advance but were cut down by what seemed to be a swarm of arrows, much like a force of locusts. Quickly, the field became a killing ground as the English archers, with apparently unerring aim, calmly stood and, with each of several thousand archers firing fifteen shots every minute, piled up bodies in heaps. Unable to get anywhere close to their tormentors, the French foot soldiers wavered and began to mill around in a confused mass, terrorized by the constant buzz of arrows that came at them with terrible speed. The muddy ground was turned into a quagmire soaked with blood.

The French knights now decided to go on the attack. Their progress was badly hampered by the disorder of the French footmen and the mud, a circumstance that made them easy targets for the longbowmen. Within minutes, amid the sounds of screaming men and horses, the knights began falling by the hundreds as the English archers cut them down, too. The French knights could hardly believe what was happening; incredibly, at ranges of 200 yards, the arrows were penetrating the finest armor as if it were paper. But the knights bravely pressed on, inspired by the example of their leader, Charles d'Albert, Chief Constable of France, who urged his knights forward as he charged headlong toward the English archers—and promptly was killed by an arrow that punched through his breast-plate armor and into his heart. He died before he could get within 100 yards of the solid ranks of English archers, who were like machines, methodically loading their bows, firing, then reloading and firing again, shooting volleys of arrows every few seconds.

Three times the surviving knights charged, and each time the archers mowed them down. A few French knights managed to penetrate the ranks of English archers, killing some of them, but they in turn were killed before they could get any farther. Finally, in fewer than ninety minutes, the slaughter ended. The French retreated, leaving nearly ten thousand dead and one thousand prisoners. The dead included virtually every French knight who participated in the battle; the flower of French knighthood was gone forever in a tangle of dead men and horses. The English had lost 113 men.

"We band of brothers," to use Shakespeare's memorable phrase, these common farmers, had humbled the world's greatest military force in what remains as one of the most stunning and lopsided victories in military history. It was a victory whose impact cannot be overstated. In fewer than two hours, a military weapons system that had remained unchallenged for more than a thousand years was swept away. Dominance of the battlefield had been restored to the foot soldier, a dominance that would remain for many centuries to come. The old feudal order died that morning at Agincourt; ordinary men had slaughtered the feudal upper class, and that class would never rise again. The social and political consequences were enormous: from that moment; ordinary men began to understand that they could be masters of their own destinies, a radical thought that over the next three centuries would result in a political sea change.

All this flowed from a triumph of science. That may seem an odd conclusion to draw from what at first glance appears to be an unimpressive-looking piece of medieval technology: a long piece of flexible wood, some string, and arrows with feathers. But look closer, for the longbow represented one of the more outstanding examples of how the human mind is able to use applied science to convert a few basic scientific principles into a tool of destruction.

• • •

No one knows exactly how the longbow was invented, or who was responsible. It is known that the longbow originated in Wales sometime during the late twelfth century; there are references to it in chronicles dating from 1188. Almost certainly, the longbow began as a hunting weapon, developed to kill game at longer distances than conventional bows, and with greater accuracy. Whatever the genesis, the longbow even today stands as a remarkable example of technical development, a perfect balance of strength and kinetic energy. Like the war engines of the ancient Greeks, the longbow represents an impressive scientific development in a time long before computers, precision instruments, and testing laboratories. It was an especially impressive achievement in twelfth-century Wales, where there is no evidence that the Welsh had anything more than a knowledge of rudimentary arithmetic; certainly, in a region of subsistence farmers and tradesmen who were almost exclusively illiterate, there was no science of any kind.

Nevertheless, they somehow developed the most brilliant (and deadly) technology of the age. In modern scientific terms, the longbow is referred to as a "self-composite weapon," meaning that it contains within its structure everything necessary to make it function at maximum efficiency. Obviously, the Welsh had no knowledge of such a concept, but there must have been an extensive research and development process to arrive at the essential first step in the longbow's development: What was the ideal construction material to achieve maximum draw power and kinetic energy? The answer was Mediterranean yew, which was discovered to be the springiest and sturdiest of all woods. Further, it had to be sapwood (the perfect natural material to resist tension) lying next to heartwood (the perfect natural material to resist compression), and kept supple with applications of oil, wax, or animal fat, representing another research and development effort to arrive at the precise combination to guarantee maximum suppleness. Additionally, the material had to be waterproof, to prevent warping of the bow in wet weather or temperature extremes; the longbow was intended as an all-weather instrument. The final oil-wax-animal fat formula was modified to use on the bowstring to keep that in perfect working order, itself a triumph of research to discover just the right combination of materials to handle the strain of draw.

Next came the design of the bow itself. Again, an impressive scientific conception lay behind the final design, a relatively thin, gently curved piece of wood up to 6 feet, 4 inches in length. Somebody figured out that there is a direct correlation between length and tensile power: the longer the bow, the longer the bowstring, which in turn means the kinetic energy increases in proportion to the length of draw on the bowstring. Another conceptual breakthrough was the realization that conventional bows, even the composite bow, were shot while aimed from the shooter's chest, which reduced accuracy. A longbow, because of its length, was aimed by drawing the bow-string to the shooter's eye, which meant archers could sight their targets along the arrow, allowing them to develop great accuracy as they adjusted for such variables as windage.

The longbow had tremendous kinetic energy and thus tremendous hitting power. Modern research in kinetic energy has indicated that some 75 foot-pounds of foot-power in a bow are necessary to propel an arrow with sufficient force to penetrate human skin, and about 150 pounds of foot-power are needed to cause serious injury. Modern tests on thirteenth-century longbows reveal that they could generate a maximum of nearly 1,400 pounds of foot-power, an astonishing amount of energy made all the more deadly by still another scientific triumph, the design of the longbow's arrows. Made out of 2 or 3 ounces of very light wood to achieve maximum flight dynamics, the arrows were 37 inches long and tipped with a cone-shaped warhead that tapered sharply to an unbarbed steel point, representing the perfect aerodynamic shape for flight and penetration of the target, a concept adopted centuries later for rifle bullets, artillery shells, and rocket nose cones.

How the poor farmers of Wales managed to arrive at such scientifically sophisticated concepts remains a mystery, but their genius didn't stop there. To make the arrows highly accurate, they developed a guidance system that used four rows of swept-back feathers on the rear end of the arrow that gave the shafts a very level, steady flight with minimum wind resistance, which meant they flew to their targets at very high speed—with the distinct, beelike buzzing sound that so terrified the French infantry at Agincourt. Almost certainly, a period of experimentation and development was necessary before arriving at the final design, which used pheasant feathers (light but strong) for the flight guidance process. In the process, the anonymous Welsh designers discovered that swept-back feathers worked best, a brilliant concept later reborn in the guidance vanes of rockets and the swept-back wings of modern jet aircraft. Further, they learned to design arrows of different weights, adaptable for different purposes—a "flight" arrow for long range, and a slightly heavier "sheaf" arrow used as a man-killer. The newly designed arrows, made to high tolerances, were models of craftsmanship and represented a major leap in technological improvement over the shorter arrows of the time, which had nowhere near the quality, range, accuracy, and hitting power of the longbow's shafts.

It is possible that the longbow would have remained simply an efficient hunting weapon had not someone recognized its immense military potential and turned it into a revolutionary weapon of war. But Edward I, king of England, did recognize that potential—and how it would solve a very serious problem.

• • •

Edward, busy subduing the Scots and other peoples in the British Isles who were not especially taken with his idea that they come under English domination in the name of creating a British kingdom, wanted to create a great national English army that would make Britain invulnerable to any outside invaders (and, possibly, allow him to do a little invading of his own). The problem, however, was cost—specifically, the huge amount of money needed to build and maintain the weapons system that was the mainstay of all military organizations: the knight.

Knights sucked gold from treasuries like no other military technology. First, unlike common infantrymen, the knight had to be a military professional devoted strictly to the state. The armored knight required constant practice and training to become proficient, so the only way to maintain a force of knights was to keep them on a royal roster and thus instantly available to suit up and go off to war. (This is why the profession of knighthood soon became restricted to aristocrats, because their inherited wealth allowed them to become knights without having to worry about supporting themselves—and drawing a state salary.)

Then there was the matter of the knights' equipment. Armor, which had to be made by hand by highly skilled craftsmen, was frightfully expensive; the armorers of Augsburg, Germany, who produced the highest quality and most advanced body armor, charged a pretty penny for their wares. And it didn't stop there: kings had to maintain staffs of their own armorers to keep the armor in repair, blacksmiths to keep the knights' horses shod, horse breeders who developed, maintained, and trained the sturdy horses for knights, and various assistants who handled assorted necessary tasks for a knight—such as lifting him onto his horse after he had suited up in armor weighing nearly 270 pounds.

All in all, a very expensive military weapon to maintain, and few kingdoms could afford the cost. Edward's kingdom didn't have that kind of cash, so he began to think of an alternative: Was there a new weapon that could be developed to end the armored knights' domination of warfare—in other words, a new wonder weapon that would restore military supremacy to the foot soldier, a much cheaper commodity?

The answer came in 1282, when Edward led an army into Wales to conquer and pacify the increasingly rebellious Welsh, who had announced their intention of forming their own kingdom. Edward didn't have a force of knights, but he did have a few modestly armed cavalrymen, considered more than enough protection against ill-armed Welsh rebels—or so he thought. One day, Edward watched as one of his cavalrymen, attempting to flush out some Welsh rebels from a grove of trees, spotted a Welshman carrying what appeared to be a longbow that was somewhat taller than himself. The cavalryman headed toward him at full gallop, at which point the Welshman quickly unlimbered his bow, strung an arrow into it, and fired. The arrow struck the cavalryman in the thigh, penetrating his leg armor, then went clean through the thigh and the saddle, killing the horse—at a range of nearly 250 yards. Several days later, ashen-faced English soldiers told Edward of an equally unsettling encounter: they were attacked that morning by a superior Welsh force armed with strange-looking tallbows. The Englishmen fled to the safety of a nearby castle, and had just managed to close a massive, 6-inch-thick oak door behind them as the Welsh archers fired from 200 yards away. The arrows struck the door, embedding themselves some 4 inches deep into the wood. Edward, examining the arrows still stuck in the door—and no doubt imagining what terrible damage they would have done had they hit human flesh—was impressed. At that moment, he found the solution to his problem.

Edward realized the Welsh had not learned to organize their long-bowmen into disciplined military units that could deliver a high volume of concentrated fire—fortunately for him, since such units would have shot his infantry army to pieces. When he finally finished subduing Wales, he set about organizing the longbowmen into state-endowed militias and training them in the art of war. In exchange for government largesse, they consented to be relentlessly drilled to stand in ranks and deliver massed fire on command; with practice, the best archers could get off fifteen shots in a minute, all with great accuracy. The archers also were drilled in hitting moving and stationary targets with equal accuracy, and to maintain their positions even in the face of onrushing armored knights or charging masses of infantry.

To further develop a proficient corps of longbowmen, Edward sponsored a series of shooting contests around the country, with the then lavish sum of 30 shillings (about $2 in today's currency) awarded to winners. Archery became the kingdom's great spectator sport, with hordes of spectators on hand to watch the greatest archers vie in hitting targets up to 400 yards away. (Among the contestants was Wales's greatest archer, the legendary Robin Hood, who was said to have won one archery shooting contest by firing his arrow into a rival shooter's shaft, which was resting in the bull's-eye of a target several hundred yards distant.)

Edward spared no expense or effort to develop his new weapon. Angering ecclesiastical authorities, he amended the Sabbath Law, which mandated only church attendance as a legally permissible activity on the Sabbath, to add one more: participation in archery contests. He did not live to see the full fruition of his efforts, but his grandson Edward III, did, eventually creating an army dominated by battalions of highly trained archers who were armed with the most effective and versatile personal weapon ever to appear on the battlefield. Now at the service of a united English kingdom, the longbowmen headed off to France to fight in the Hundred Years War, where they would change the course of history.

It is a tribute to the brilliance of the man or men who devised the longbow that in the ninety-plus years required for Edward I and his successors to fully militarize it, there were no improvements whatsoever made to what some Englishmen took to calling "the great equalizer." That's because no improvements were necessary: in one of the rare examples of technical perfection in the history of technology, the longbow had achieved the highest possible level of its technical development under then-prevailing technology. The longbows that ended knighthood at Agincourt were virtually the same longbows that so impressed Edward I when he first saw them in action more than a century before. (As late as 1940, there were some authorities in England who seriously proposed the idea of reviving use of the longbow, unchanged since Agincourt, to help stave off an anticipated German invasion.)

Everybody now wanted these new miracle weapons, but they immediately discovered that although dominant on the battlefield, the longbows carried a serious drawback: Who would shoot them? Unlike such basic and unsophisticated infantry weapons as the sword and battle-ax, for which anyone could acquire basic proficiency pretty quickly, mastering the longbow was not something that could be achieved in a short while. As both Edward I and his grandson learned, longbows needed good longbowmen to operate at peak efficiency, and that was a commodity attainable only by long periods of training and many years of constant practice. Some ten to twenty years of devoted training were required before a longbowman could attain complete proficiency with the weapon—when he could confidently stand his ground, unarmored and unprotected save for his bow, and rapidly fire arrows at onrushing knights or infantry soldiers, taking less than a few seconds to aim and fire at a target.

There was an even more serious problem the longbows created, a problem that arose in the very nature of medieval Europe. Up to that point, there had been a well-established pattern: a wonder weapon of one kind or another would be developed, then science would devise a counterweapon, which in turn would be trumped by another wonder weapon, and so on. But in the wreckage of ancient Greece and Rome, the destruction of the Alexandria *museion*, and the disappearance or scattering of the classic scientific texts, science had gone to sleep. The barbarian tribes that swept into western Europe had no scientific tradition, their development of armored cavalry notwithstanding. When they finally settled down, they created a series of vassal states, basically subsistence and nonmonetized economies that became feudal orders. It was an age of chaos and fear, and science languished in a time of myth, legend, omens, demons, magic, astrology, divinations, and sorcery. This time of unending war and destruction became known as the Dark Ages, when these small states constantly warred with each other. Combined with the threats that existed on all sides—Viking raiders, Islamic incursions, attacks by the Mongols from the East, assaults by tribesmen from the Russian steppe—war was omnipresent. "It is the last hour," Gregory the Great, one of the feudal lords, wrote morosely to the patriarch of Constantinople in 595 C.E. "Pestilence and sword are raging in the world. Nation is rising against nation, the whole fabric of things is being shaken."

This dark assessment represented only a slight exaggeration, for indeed a terrible darkness had settled over the Western world, unlit by even a flicker of science—except for some bustling armories, where the science of metallurgy was in full flower, busy trying to perfect what every feudal ruler knew was the ultimate weapon that would never be overthrown, the armored knight.

• • •

Feudal minds found it impossible to conceive of any real threat to the armored man on horseback, except for another, better, man on horseback. This ultimate weapon had come about because of a great conceptual breakthrough. Eons before, man had conceived the idea of putting soldiers on horseback, but right through the time of the Roman Empire, they rode bareback, using their knees to keep themselves mounted. Difficult enough to do, and even more difficult when the cavalryman, one hand occupied with the bridle, tried to use the other to swing a weapon, especially at full gallop. And using a bow and arrow with any kind of accuracy from so precarious a mount was a real challenge. Moreover, disciplined infantry learned that a cavalryman could be pulled off a horse.

No infantryman could pull an armored cavalryman off his horse, thanks to the Goth saddle and stirrup system that shattered the Roman infantry and transformed horse and man into the ultimate weapon. Since nobody was able to conceive of a better one, all effort was devoted to improving it. In 732 C.E. Charles Martel, leader of the Franks, introduced formations of what he called *cataphracts* (later "knights"), cavalrymen protected by a full suit of body armor and armed with long, thick lances. Later, Charlemagne created a force of armored knights who were trained with a new breed of very strong European workhorses that had backbones up to 7 feet off the ground. Trained from colthood as warhorses, these large horses, combined with an armored knight firmly saddled, created a shock weapon that even the most disciplined infantry found difficult to resist. Until the advent of the longbow, infantrymen had little to counteract the mo-

bility and power of a full-scale cavalry charge, when hundreds of
armored horsemen, invulnerable to arrows, would smash into an infantry formation at full gallop, skewering men with lances, or striking them down with maces, battle-axes, and long swords.

Events seemed to support the conviction that armored cavalry
represented an ultimate weapon. In one of the age's most decisive
battles, Hastings in 1066, a Saxon army composed only of infantry
armed with shields, iron-tipped spears, throwing axes, swords, and
bows and arrows, confronted an invading army of Normans whose
forces included eight thousand armored cavalry. Norman archers
fired a volley of arrows, against which the Saxons adopted the tactic
borrowed from the ancient Greeks at Marathon, raising their shields
above their heads in a "turtle" defense. It was the moment the
Norman cavalry was waiting for: they swept in at full gallop, crashing into the Saxon infantrymen with lances. The Saxons crumbled
under the onslaught, and with only leather caps and shirts to protect
them, they were cut to pieces. By the end of the day, the Normans
were masters of England.

There was more evidence, if needed, that the armored knight was
the master of the battlefield. Science made that weapon ever more
powerful—or, more accurately, one particular scientific discipline:
metallurgy. Indeed, the obsession to improve the armored knight
created modern metallurgical science, for it was the key to technical
improvement. It also created one of history's most ruinous arms
races as feudal states sank vast sums into efforts that sought better,
thicker, more pliable, and more functional armor, along with more
powerful hand weapons for the knights. There were considerable
ripple effects, especially the strong strides made in improving mining techniques to satisfy the growing demand for ferrous ores, and
the creation of the entirely new technical classes of armorers and
metalsmiths, lavishly paid by their feudal patrons.

The center of mining and armor technology was Augsburg, in
Germany, and that was no coincidence. Augsburg was near one
of Europe's major deposits of iron ore, and the demand for metal
from feudal states building forces of armored knights soon created
a booming mining industry and an equally flourishing armorer
business. To the annoyance of their customers throughout feudal
Europe, the Germans charged sky-high prices, aware that those customers had no alternative: German armor was the best in the world,

and if a customer didn't like the prices, he could sally forth on his next war with sticks and stones. Underwritten by these lavish profits, the German armorers could afford an extensive research and development effort. It resulted in high-temperature blast ovens; new smelting techniques; lighter, but stronger, armor; plate armor (overlapping sheets of metal that protected knights from battle-axes); steel helmets with movable visors that covered the entire head; suits of armor with supple joints; razor-sharp, double-edged swords made of steel so hard that even the hardest blows wouldn't dent them—and those two essential components of modern technology, the screw and the rivet.

Because of this scientific research effort, the armorers were able to fulfill the insatiable demand for improved weapons and armor (for a price). It also allowed them to concoct a quick technological fix when the armored knight encountered his first serious threat, the crossbow.

• • •

Essentially a handheld *ballista,* the crossbow originated in ancient Greece, where it never achieved anything near the status of a war-winning weapon. Its chief drawback was a slow rate of fire, a problem that remained insoluble. The crossbow fired a steel bolt by means of a drawstring that had to be drawn back, locked into position, then released by a trigger. A good deal of strength was required for this process. In the early eleventh century, a group of Italian engineers, aware of the potential riches in a weapon that could defeat the armored knights, hit upon the idea of reviving the Greek crossbow, with a few technical improvements. The chief one was a mechanical winch system that the firer used to cock the weapon, a device that avoided the necessity for a lot of musclepower. An infantryman would rest the stock of the weapon on the ground for support, then winch the drawstring into firing position. The crossbow would then be loaded with a pointed steel bolt that was fitted into a slot, and fired with a trigger that would release the bowstring, which was under great tension from the machine's two arms. The release snapped the bowstring forward, releasing a tremendous amount of kinetic energy that could project the bolt up to 500 yards away at high speed.

Initially, the crossbow showed great promise of becoming a new wonder weapon, but its limitations soon became obvious. For one thing, the crudely made steel bolts it fired were not aerodynamically designed, so they tended to vary wildly in accuracy. For another, the winching mechanism, while ingenious, did not solve the weapon's basic flaw, a slow rate of fire. Even the most highly trained crossbowmen were able to get off only two shots a minute, a not very helpful rate of fire in the event of a horde of armored horsemen bearing down on an infantry formation. And an infantryman in the process of winching his weapon was about as vulnerable a target as could be imagined. Finally, crossbows were very expensive weapons. To make them flexible and at the same time strong, a composite material was needed; the best turned out to be buffalo horns and wood. There was plenty of wood around, but buffalo horns were in short supply, so that turned out to be an expensive component. Worse, the buffalo-horn-and-wood composite had to be glued together with a glue that would not loosen because of weather conditions, heat extremes, or the considerable strain induced by all that winching and firing. The Italians experimented with various glues, finally emerging with a superglue that, unfortunately for their pocketbooks, was made from the skin of the roof of the mouth of the Volga sturgeon. That ignited a sudden rush for Volga sturgeon, and poor Russian fishermen, who considered themselves lucky to earn a few kopeks for their catches, now discovered they were sitting on a gold mine—and jacked up prices accordingly.

Nevertheless, the crossbow posed a real threat to knights armored only in chain mail (linked rings of metal, very popular because it was the cheapest kind of armor to produce). A crossbow bolt weighing 3 ounces easily penetrated chain mail, setting off a drive to come up with a counter. It resulted in the wizards of Augsburg inventing plate armor, sheets of overlapping steel plate put over chain mail, a brilliant design that made armor somewhat flexible, thus absorbing much of the crossbow bolt's shock power by actually "giving," then flattening to halt penetration. They also devised armor for the horse, which blocked a favorite tactic of crossbowmen: shooting a knight's horse out from under him, which allowed other infantrymen to hack the immobile knight to death. The new armor effectively stopped crossbow bolts, except for those fired at very close ranges, which didn't prevent a concerned Second Lateran Council in 1139 from forbidding the use of the "un-Christian" crossbow in war "under penalty of anathema" (adding, however, that the weapon could be used against such "heathens" as Muslims).

The metallurgists had triumphed again, but elsewhere in Europe, there were no equivalent scientific-technical races for the simple reason that there was no science. Feudal lords walled themselves in elaborate castles meant to be a self-contained, invulnerable world— very high walls up to 18 feet thick, round towers, parapets, water-filled moats, drawbridges, and a heavy iron gate virtually immune to ramming. The castles were really a series of two or three castles, one inside the other. The inner walls were taller than the outer walls, allowing troops on those walls to prevent a besieging enemy from using the outer walls as a springboard for further attacks against a castle's inner defenses. Castles were defended by several successive lines of retreat, with each line designed to extract maximum casualties from an attacking force, whittling it to impotency before it could reach the castle's inner nerve center.

Into these castles the feudal lords gathered their courts and knights and administered their kingdoms. Outside the walls lived the lower feudal orders, including serfs and vassals, guaranteed protection inside the castle in the event of attack. Actually, castles stood very little threat from attackers; few were ever captured except by surprise or treachery. And that was because the art of siege-craft had remained virtually unchanged from the time of the ancient Greek siege machines many centuries before. There was no science to develop new methods of overcoming castle defenses, so besiegers attacked castles the same way the ancient Greeks attacked each other's fortresses. The besiegers would bring up siege machines containing soldiers, attempting to get close to the walls, as other troops, using "turtles," moved to the base of the walls and began to climb them on scaling ladders—all the while trying to avoid death from arrows shot from parapets, big rocks dropped on their heads, and boiling oil poured on them.

Meanwhile, catapults and *ballistae*, virtually unchanged from Roman times, would be brought up to add firepower. As in ancient warfare, the catapults heaved huge rocks over the defensive walls against softer targets, while the *ballistae*, armed with steel bolts, tried to pick off defenders on the walls. (To be sure, the Europeans added a few refinements, such as launching putrid corpses in an attempt to spread deadly disease, or, in the case of a siege against a German castle, two thousand cartloads of manure).

• • •

While the Europeans were fighting wars in ways that would not have astonished Archimedes or Julius Caesar, a brief, sudden flash of science many miles to the east managed to save the last vestige of the Roman Empire—for a while, anyway. It happened in Byzantium, the state created by Emperor Constantine when he converted to Christianity in 324 C.E. and moved the eastern half of the Roman Empire to the Greek colony of Byzantium, in Turkey. The Byzantine Empire's capital, Constantinople, became a great trading hub, but it faced danger from a host of covetous enemies, especially assorted Muslim satrapies.

Byzantium had a solid corps of scientists, most of them Greek, into which it tapped for some sort of military technology that would make it invulnerable. The Greeks succeeded beyond anyone's wildest hopes, for they were responsible for that age's thermonuclear weapon, "Greek fire." In fact, Greek fire wasn't entirely new. In about 400 B.C.E., scientists in ancient Greece, called upon to devise a new weapon that would break the will of besieged fortresses, came up with what they called a "fire weapon." A product of early Greek advances in chemistry, it combined pitch, sulfur, granulated frankincense, and pine sawdust in a mixture that was packed into cloth sacks. Set alight, the sacks would be launched by catapults into the interior of fortresses, where they spread flames in all directions on contact. To the distress of defenders, pouring water on the flames seemed only to spread them farther.

In 673 C.E. the Byzantines were desperate for a weapon that would protect Constantinople, a port city, from attack by sea. The Byzantine navy was weak, so there was a critical need for some kind of weapon that would destroy invasion fleets and deter any landward attack. The Greek scientists decided that a fire weapon would be especially effective against wooden ships, but a careful look at the "fire weapon" of a thousand years before instantly revealed its limitations. Catapults were not that accurate, so the prospect of a machine with sufficient accuracy to launch a sack—a not very aerodynamic missile, in any event—from hundreds of yards away and hit a moving ship was remote at best. Launching such weapons from a ship was a possibility, but there again, the problem of accuracy arose: How to hit a target from a ship rolling in ocean waves?

At this point the Greek scientists recalled that in their travels around the Middle East, they had seen a strange, smelly black substance oozing out of the ground. Nobody knew quite what to do with the substance (the Greeks called it "petroleum"), but the scientists noted that the black goo was highly flammable, whereupon they made the conceptual connection: convert "Greek fire" into a liquid weapon. In short order they devised an amazing weapon that anticipated the flamethrowers and napalm bombs of many centuries later—a mixture of petroleum and calcium phosphide (made from lime, bones, and urine) that amounted to a more powerful, liquefied version of the ancient "Greek fire." Then they designed a weapons delivery system for it, an apparatus consisting of a large storage tank containing the petroleum mixture, a hand-turned pump, a hose made of animal hide, and a nozzle with a spark-driven "trigger." The next step was modifying small galleys to carry the new weapon; the tank and the pump were put belowdecks, while the hose and the nozzle were fitted abovedecks, on the bow.

The weapon came just in time, because the Saracens launched an invasion fleet toward Constantinople, intending to take the city. The fleet was met by several small, fast Byzantine galleys at sea. Using the superior maneuverability of their smaller ships, the Byzantines swung close to the Saracen ships as a crew belowdecks pumped furiously and a man wielding the nozzle, protected by archers, aimed it at a target. When the liquid reached the nozzle, he struck two pieces of flint together, producing a spark that converted the liquid into a spurt of liquid fire that arched out as far as 50 yards. One Saracen ship, then another, was engulfed in flames; the rest fled in terror.

Word soon spread of this miracle weapon, and traders with an eye on the immense profit to be made should they discover its formula besieged every Byzantine business contact they knew: What would it cost to obtain it? But the formula remained Byzantium's most closely guarded secret; those attempting to find out were told angels had conveyed the formula directly from God to Emperor Constantine. Not many people believed this fairy tale, but all attempts to uncover the formula failed. Among the most avid formula-seekers were the Ottoman Turks, determined to seize Constantinople, the city they believed properly belonged in their sphere of influence. So long as the Byzantines possessed the ultimate weapon, however, there was no way to accomplish that militarily—as events of 716 C.E. proved, when nearly all of an entire fleet of Ottoman Turkish ships attempting to carry out an invasion of Byzantium was destroyed by Byzantine ships spewing Greek fire.

The Ottoman Turks never did discover the secret of Greek fire, but no one else did, either. In the end, however, it hardly mattered. The Byzantine Empire, convinced it was invulnerable because it possessed an ultimate weapon that would never be defeated, grew complacent. It never tapped again into science to keep its military edge. It was only a question of time before relentless Arab pressure against a complacent Byzantine state began to tell—especially after the Byzantines, believing that their Greek fire would allow them to ignore any foreign threats, focused inward on their political and religious schisms and grew progressively weaker. Not even the wonder weapon of Greek fire could paper over such cracks, and by 971 C.E., Byzantium's time of glory was over. It would finally collapse under renewed assault by a resurgent Islam; Constantinople eventually fell and was renamed Istanbul.

• • •

The early sequence of events in Byzantium passed unnoticed by the European kings, cooped up in their castles and isolated from the rest of the world as they spent most of their time preoccupied with trying to survive in the sprawling patchwork of kingdoms and an environment of constant war. That suddenly changed in 1071, when Arab armies inflicted a catastrophic defeat on Byzantium, which, as a result, lost almost all of the eastern part of its empire. The defeat signaled a real danger from a resurgent Islam, whose forces occupied part of Spain and had swept into the Holy Land. The resurgence had begun in 632 C.E., when Muhammad the Prophet died and was replaced by Abu Bakr, who announced the "flame of Islam," by which he meant that Islam would embark on a drive of conquest to spread its doctrine everywhere—including Europe. As alarm bells rang all over Europe, two critically important events took place: the ecclesiastical establishment in Europe called on its kings to provide military forces for a "holy crusade" to expel Islam from the Holy Land,

and the rulers of the Spanish kingdoms decided they would drive Islam out of Spain.

Coincidentally, while these two events began to unfold, a group of ecclesiastical scholars in France discovered that they shared a heretical thought: nature is intelligible because the human mind and nature proceed according to the same rational laws. Therefore, the answers to nature's mysteries were not to be found in Scripture, but in rational thought and investigation—in other words, science. They gathered themselves into an informal university called the School of Chartres, then begin trying to divine the great mysteries, only to discover they were missing essential texts, the classic works of philosophy and science of the ancient Greeks.

Within a few years, the scholars would get those texts. In the process, the greatest scientific revolution in history would be ignited, with consequences not all to the good.

• • •

The main credit for that revolution accrues to an enlightened Spanish king named Alfonso VII, ruler of the Spanish kingdoms of Castile and Leon. Alfonso, whose army led the Spanish assault against Islam's grip on southern Spain, issued strict orders to his troops: there would be none of the standard pillaging, looting, sacking, slaughtering, and destroying when they took over Islamic-controlled cities. These cities, Alfonso decreed, contained "important things" that should be studied by European scholars for what might be of "possible value."

That turned out to be an understatement. In 1085, when Toledo fell to Alfonso's troops and the Muslims fled, they left behind a magic city whose wonders caused the Spanish troops to stand awestruck in the streets, hardly able to believe what they were seeing—beauty salons, rich meals divided into courses, seasonal dress, drinking glasses, rhythmic music, regular postal service, paper money, and lush fountains in the entrance courtyards of houses. There were other treasures of Toledo whose significance was not apparent to the soldiers, but it was to the scholars who now made their way across the Pyrenees into Spain to feast their minds upon abacuses; astrolabes (navigation instruments that used star positions for orientation); algebra; trigonometry; and an odd-looking yet elegantly simple number system of 0, 1, 2, 3, 4, 5, 6, 7, 8, 9, which the Muslims had borrowed from the Hindus (and which would in turn be adopted by the West).

Best of all, there were libraries stuffed full of manuscripts. And what manuscripts: the Muslims, as it turned out, had painstakingly collected and translated the scattered intellectual treasures from the wreckage of ancient Greece—the works of Plato, Aristotle, Archimedes, Pythagoras, Euclid. Other manuscripts were falling into the hands of the Crusaders, who experienced their own culture shock. When they swarmed into Constantinople in an attempt to wrest it from the Muslims, they encountered a glittering world they could not have imagined, especially the food. They came from a world of drafty, dirty castles and a diet of often-rotting meat and fish in the summer (since there was no way to keep them fresh), followed by a winter diet of old salted meat, fish, and a few roots—a cuisine so dreadful it was called "that which goes with bread." In Constantinople the Crusaders saw people actually enjoying their food (often at curious public eating places called restaurants), largely because they could season it with strange powders the Crusaders had never seen: pepper, sugar, cinnamon, ginger, cloves, and saffron. The Crusaders shipped back to Europe samples of these spices, setting off a European obsession for these marvelous powders, with later consequences for the course of world history.

Meanwhile, the Crusaders were also busily shipping the contents of plundered Muslim libraries. Greek science was now returning west, and like opening the windows of a musty room, this accumulated wisdom, translated into Latin, the scholarly language of the day, blew a fresh breeze into Europe. It was not entirely welcomed; the mighty power of the papacy worried that this sudden flood of knowledge was dangerous, since there was no way of knowing how many "erroneous" ideas might be lurking in all those manuscripts, sufficient to undermine faith. The great intellectual explosion might have ended right there, except for a monk named Thomas Aquinas, the church's greatest thinker, who neatly solved the problem with one of the West's first great philosophical treatises. He bridged the gap between faith and reason by deducing that there were two kinds of knowledge: one related to revelation, the province of theology, and the other to the natural world, the province of reason and philosophy. In a famous phrase that was to set in motion the spirit of free inquiry that resulted in the scientific revolution he said, "God cannot make the sum of internal angles of a triangle add up to more than two right angles." Reason, he concluded in another memorable phrase, was the "bride of faith."

But another religious philosopher, the English cleric and polymath Roger Bacon, while he shared Aquinas's view, looked deeper and saw a darker vision. He predicted that the tidal wave of Greek science texts flooding into Europe would cause a scientific revolution, something that had the potential to transform human thought and lift humans from barbarism and superstition. However, given the nature of man, Bacon feared the new knowledge would be used primarily to devise new ways for men to kill each other. In an extraordinary vision, he predicted a future of misuse of science that would produce wondrous war weapons—submarines, "flying ships," armored vehicles, propelled ships, cannons—and great wars in which men would slay each other in great numbers. Bacon added an even darker prediction: someday, morality would be subordinated to science.

Unfortunately, he turned out to be absolutely right.

• • •

Bacon and other thinkers toyed with the idea of keeping the new knowledge restricted to a small circle of scholars who could be depended on to keep it from the kind of men who would use science only to devise new and better ways of killing other men. For all his talents as a prophet, however, Bacon did not foresee that a German goldsmith named Johann Gutenberg would make such restriction impossible. Gutenberg's idea in 1439 for movable type, which would create the final impetus for the scientific revolution, made all those ancient texts accessible to everyone because the new printing process could make an unlimited number of copies. Further, scientists could disseminate their discoveries and theories widely by publishing them in books free of copying errors, since each copy was a precise replica of the original.

Bacon's uneasiness was validated, to a certain extent, by what happened as the revolution of the printing press exploded across Europe: the major impetus for the spread of printing was military. European soldiers and kings searching for insights on such matters as how to organize a first-rate military force or handy tips on how to knock down a castle wall wanted to delve into the ancient texts about war, science, and military strategy, hoping the great military and scientific minds of the ancient world had some answers. The effort posthumously revived the career of the long-forgotten Flavius Vegetius Renatus, a Roman patrician who in 378 C.E. made a careful analysis of why the Roman legions had collapsed in the face of the barbarian invasions. (Basically, he concluded that the Romans needed another Julius Caesar, who, regrettably, had been dead for more than three hundred years and was therefore unavailable.) Vegetius collected his thoughts on the tactics, strategy, weapons, and generalship of the Roman military, along with a detailed analysis of why the legions dominated the world for so long, in a lengthy manuscript titled *De Rei Militari (On Military Affairs)*. It amounted to a thinking man's how-to guide on building, maintaining, and leading a great military organization, a blueprint for any budding militarist seeking to create a war machine.

The manuscript was among a large cache of ancient writings seized by the Crusaders in formerly Roman-occupied areas of Egypt and Palestine. A Venetian printer named Aldus Manutius got the bright idea that there was a market for this sort of thing, and went heavily into the military publishing business, running off ancient military texts after first translating them into Latin. Further, aware he was dealing with military men who were usually not the beneficiaries of a classic education, he prepared the books in simple forms and styles to make them more readable (including the invention of italic type). Manutius prepared an edition of *De Rei Militari*, first changing the title to *On the Military Institutions of the Romans*, a marketing strategy he thought would attract more readers hoping to learn the secrets of Roman military success. He immediately hit paydirt: the book was a sensation, and Manutius' busy shop could hardly print copies fast enough to meet demand. By 1500 some ninety-nine editions of the book had been published, the first bestseller in history. Manutius went on to produce a number of other military best-sellers, including studies of Roman military tactics by Polybius, and Greek works on the science of torsion catapults.

The diffusion of military knowledge through the printing press was an essential precondition for what has since been termed the great "military revolution" that swept across Europe. It was essentially a revolution in thinking; kings and generals learned from the new books of military science that to create military power, they would have to organize and systematize it. Logistical resources would have to be carefully organized; the plans for campaigns thought out beforehand; successes and failures objectively analyzed for "lessons learned"; and, most important, scientific and technical resources needed to be developed to keep a fighting edge. Fundamentally, the Europeans learned that war was a science—and it was science that would enable them to triumph.

This revolution in thought would eventually make the Europeans the military masters of the planet, but it also raises a question: Why didn't the Muslims, who had these ancient texts in their hands for centuries, do the same thing? The answer has everything to do with the way two different cultures perceived knowledge.

At the very moment Aquinas was telling his fellow Europeans how faith and reason could coexist, his Iranian counterpart, the leading Arab philosopher Ghazzali, concluded that the treasure of ancient texts represented social dynamite. The study of science and philosophy, he wrote, was harmful because it would shake man's faith in God and undermine the Muslim religion. Accordingly, the ruling caliph of Baghdad, to demonstrate his piety, ordered the burning of all manuscripts in the city's great library. Muslim scientists, tightly controlled by the ruling theocracies, had access to the ancient science texts, but were ordered to use any of the knowledge they gained only for religious purposes. So Muslim astronomers were told to use their knowledge only to determine the proper times of prayers for the faithful and the precise direction to face Mecca. Arab science was confined in a straitjacket of religious orthodoxy, which decreed that truth could be found only in the Koran. The ancient Greek scientific idea of logical argument and syllogistic reasoning was foreclosed to Muslim science because it challenged the immutable word of the Koran. And all science was to be tightly held within the ruling circle and never disseminated, for fear of undermining faith. As a result, the Muslim military organizations had no access to science and thus developed no new weapons or techniques.

That would prove costly to the Muslims, for as Muslim science sat still, the Europeans surged forward in a scientific explosion that eventually would destroy the civilization from which they derived so much. That explosion required only a single spark to ignite—a spark that arrived one day in Roger Bacon's study.

• • •

Sometime in 1248, Bacon was given an odd-looking contraption that had been brought back to England by one of the first missionaries to visit China. The contraption was called a "firecracker" by the Chinese, the missionary reported, and was used as a noisemaker during holiday celebrations. Among Bacon's wide-ranging scientific talents was a knowledge of chemistry, which caused him to become gravely unsettled when he took the firecracker apart in an attempt to understand why it exploded when exposed to flame.

Analyzing the black powder inside, Bacon realized that the explosion was caused by a mixture of saltpeter and other chemicals. With his farseeing vision, he also realized he was looking at something with a potential for mischief that dwarfed anything mankind had ever seen. As an early adherent of the scientific method, Bacon decided to write up his findings in a book titled *Epistolae de Secretis Operibus Artis et Nature et de Nullitate Magiae,* one of the seminal works in early science. But its key section, the formula for what would become known as gunpowder, was hidden in a cryptogram; Bacon believed that enciphering so valuable a secret somehow would keep it from being misused. As he explained, "The crowd is unable to digest scientific facts, which it scorns and misuses to its own detriment, and that of the wise. Let not pearls, then, be thrown to swine."

It amounted to an eerie preview of the attempt, many hundreds of years later, to keep the science of nuclear weapons secret—and was equally doomed to failure. As Bacon did not understand, there was no way that a scientific development that promised military supremacy could be kept secret. Some way, somehow, the secret would be uncovered, a lesson the longbowmen of England learned the hard way.

The longbow, which had revolutionized war, remained supreme on the battlefield—until just thirty-five years after Agincourt, when the inevitable occurred: somebody came up with a better weapon. As usual, the better weapon was the result of a desperate urgency; in this case, the French, determined to expel the English invaders from France, bent every effort to come up with a counter to the longbow.

Science finally provided it, and in 1450, at the village of Montigny in northern France, the reign of the longbow as the supreme weapon came to an end in one bloody afternoon.

The English, as usual, had drawn up for battle with their formidable longbowmen in front, ready to cut down attacking French infantry. But the French did not attack; instead, they stayed about 100 yards away from the English lines and set up what appeared to be large, bell-shaped jars made of iron, into which they poured a black powder, followed by stones. The mixture was tamped down into the bell-shaped area, then lit with a long match. There was an ear-shattering explosion, followed by a sharp whistling sound as hundreds of stones flew at high velocity into the English lines. In minutes, 3,774 of the 4,500 longbowmen were dead, having never fired a single arrow.

The English occupation of France, which had lasted nearly a hundred years, was at an end, but the consequences of that afternoon at Montigny extended much farther. Despite Bacon's efforts to keep science in a bottle, the genie had escaped. The English cleric's darkest visions were about to come true.

3

The Dragon's Teeth

*Blessed be those happy ages, that were strangers to
the dreadful fury of those devilish instruments of
artillery, whose inventor, I am satisfied, is now in
hell, receiving the reward of his cursed invention.*

—CERVANTES, *Don Quixote*

With the impact of a mighty thunderclap, the shocking news
rumbled into every corner of Italy that spring of 1494, news
so unsettling, so cataclysmic, that few dared believe it. By royal
couriers on relays of fast horses, by itinerant peddlers, by monks
traveling among monasteries, the news arrived in the architectural
gem of Florence, the beauty of Venice, the gaiety of Verona, the ancient grandeur of Rome, the bustling commercial center of Milan.
And these crown jewels of Italian civilization trembled in fear.

Something terrible had happened in Naples, an event that endan-
gered every city-state on the Italian peninsula. It was not simply
that the mighty fortress of Monte San Giovanni had fallen to an
invading French army—that was bad enough—it was *how* it had
fallen. As everyone knew, Monte San Giovanni, which protected the
land approaches to the city-state of Naples, was among the most
powerful such edifices in all Europe, and provably impregnable: high
walls several feet thick and a state-of-the-art defensive system that
no besieger had overcome in more than a hundred years of trying.
Only two years before, the fortress had withstood a siege lasting
seven years.

But, unbelievably, Monte San Giovanni had fallen to the French
in fewer than eight hours. The French hadn't used siege engines and
catapults, nor had they carried out the arduous and time-consuming

siege techniques that European armies had been utilizing for hundreds of years. Instead, they had wheeled thirty-six new weapons to within 150 yards of the fortress walls. Those weapons belched smoke and fire, hurling 50-pound iron balls against the walls, reducing them to rubble and killing most of the defenders as the fortress collapsed. Not one of the French besiegers suffered as much as a scratch.

And so, in a few hours on a balmy spring morning, a revolution in warfare arrived in Italy. It arrived in the form of something new and terrible—a few dozen cannons, weapons of unprecedented power that threatened to sweep all before them. At the precise moment a French cannoneer touched a long match to the fire hole of one of those new wonder weapons, he set off an explosion that changed everything.

• • •

The shock of what happened at Monte San Giovanni severely unsettled Italy, a patchwork of disunited secular city-states, papal states, and petty fiefdoms that had been warring with each other since the fall of the Roman Empire. Italy also was a battleground for the ambitions of larger kingdoms, a theater where French, Hapsburg, and Spanish dynasties enacted their dramas of aggrandizement, conquest, and power rivalry. The little Italian entities had managed to avoid extermination through adroit political maneuvering, playing off one power against another, and, when that failed, frustrating an invader with the best fortifications that money could buy (or, in the case of Venice, a 360-degree water barrier). But when the army of Charles VIII of France, fifty thousand strong, swept into Italy in 1494 on a plan of conquest, to the accompanying rumble of wheeled cannons—those wonder weapons that reduced the mightiest fortresses and castles to piles of dust and broken stone—it was clear that the old methods would no longer do. It also was clear that some kind of counter to Charles's wonder weapon would have to be devised—and fast.

Despite their political disunity, the best minds of fifteenth-century Italy uniformly bent to the task of preventing total disaster from Charles's relentless march (several small Italian principalities surrendered to him after merely seeing the approach of his cannons). In Florence, the city's artists prevailed upon their great German contemporary, Albrecht Dürer, in the city to study with the great Florentine painters, to put his brushes aside and sketch ideas for radically new fortification designs that would balk cannonballs. In Rome, Michelangelo interrupted his work on new building designs to think up ideas for various man-made barriers that might hinder Charles's army from moving along the roads. And in Milan, the most interesting mind in all Italy put his considerable brainpower to the task.

Leonardo da Vinci, a military engineer, had arrived in Milan eleven years before, looking for a new job. He was a man of many talents, but the best employment opportunities in fifteenth-century Italy, not surprisingly, were connected with military talent. In a time of constant warfare between Italian city-states, the rulers of these small kingdoms were willing to lavish parts of their considerable wealth on the kind of scientific brainpower that would make them militarily invulnerable. Aware of that, da Vinci submitted his résumé to Milan's ruler, Duke Lodovico Sforza, which emphasized his claim to be a "master of military arts." He went on to say that if employed, he would reveal to Sforza his "own secrets," which he listed as covered war wagons (an early version of the tank), portable cannons, a machine to drain water from the moats that protected castles, and a system to make ships invulnerable to cannons.

Sforza was unimpressed. What he needed were practical, available weapons and devices he could use now, not the kind of pie-in-the-sky inventions that da Vinci *might* be able to develop. He turned da Vinci down flat, indicating no interest in the applicant's mention of "certain peaceful pursuits" he could perform in the event the duke had no need of his claimed mastery of military science. Among them, da Vinci noted in one of the great understatements of all time, was "painting . . . which . . . I can do as well as any other. . . ."

Nevertheless, da Vinci decided to stay on in Milan, occupying himself with various tasks, mostly painting masterpieces, which he put aside in 1494 when the great crisis struck. For inspiration he delved into his notebooks, where for some years, during his spare time, he had been writing assorted scientific speculations, along with sketches of a wish list of futuristic weapons that sprang from his agile mind—poison arrows, scythed chariots, mortars, small-arms cartridges, air guns, steam catapults, rocket launchers, helicopters, armored vehicles, and poison gas. With striking prescience, he saw far ahead, into the weapons of destruction of the twentieth century, their practicality limited only by the lack of the kind of science and technology that would have made such weapons possible in his own time.

But in what also would be a preview of things to come, the man described by Freud as someone who "awoke too early in the darkness" underwent something of a crisis of conscience over what he was sketching on those pages. The notebooks were filled with a sense of foreboding and conflict, presaging modern science's web of guilt and responsibility. He often called war *"pazzia bestialissima"* (bestial insanity), but sketched designs for weapons and devices that would make it even more bestial. To solve the moral dilemma in which da Vinci found himself, he tried to have it both ways, describing his idea for a submarine that would sweep navies from the seas, but then refusing to reveal further details "on account of the evil nature of man." Such inventions, he wrote, would be "too satanic" to be placed into the hands of "satanic men." (Many of his notebook entries were written backward, apparently as a means of preventing anyone looking over his shoulder as he was writing and acquire that "satanic" knowledge.) Despite his uneasiness, da Vinci later: would work as a military engineer when he left Milan and went on to Venice, where he was hired by the city-state's government-run arsenal, and where, among other things, he developed new ways to cast cannons. Before leaving, he threw himself into the task of trying to come up with some sort of defense to the grave threat faced by Milan—and every other Italian city-state—from the French invasion. There is no record to indicate whether da Vinci, either by himself or in concert with others, came up with the innovative idea that finally brought the French to a halt. But it can be safely surmised that as a man with a scientific mind, he understood that the solution to the problem had to involve science of some kind. After all, science had created the problem in the first place.

• • •

The new French wonder weapon was born at Agincourt in 1415, when the English longbow destroyed knighthood and suddenly provided England with an ultimate weapon that threatened to make their occupation of northern France permanent—and, possibly, permit even greater inroads on French soil. Charles VII, the then reigning French king, vowed to expel the English from France. To that end, he decided an entirely new approach would be necessary. Among Europe's first great soldier-kings, Charles was a well-read, cultured man who had eagerly devoured the ancient military classics. And that reading convinced him that only an all-out, organized scientific-technical effort would be able to come up with a solution to the longbow crisis. Borrowing freely from concepts first invented by the Assyrians, Dionysus, and Alexander the Great, Charles recruited the best scientists, technicians, and engineers he could find on the Continent—German armorers, French scholar-scientists, Italian mathematicians, Europe's best metallurgists, Czech alchemists—with lavish offers of gold and organized them into a full-scale military research and development program. Their marching orders were simple: develop a new ultimate weapon to defeat the longbow, and do it fast.

Charles's scientific brain trust did not have to look far for a solution, because as things turned out, the ultimate weapon they were seeking was already at hand. But the problem was that nobody had yet figured out how to develop its full potential.

• • •

The origins of gunpowder are obscure, but it is believed to have been invented in China in about 900 C.E. By 1150 the Chinese were using the volatile substance to power small rockets and as the chief ingredient to make firecrackers pop. It was still something of a toy until the early thirteenth century, when the threat of Mongol invasion inspired a Chinese effort to come up with some sort of weapon that would decimate the hordes of Mongol cavalry. There is no record of who made the next critical step, turning gunpowder's power into a propellant for an antipersonnel weapon. The Mongols soon learned the power of that new terror weapon, tubes of bamboo or iron mounted on the shafts of lances, which contained a mix of gunpowder and stones, lead pellets, and pottery fragments. When a Chinese soldier touched a match to the gunpowder, the world's first guns spewed their ammunition forward in a scatter pattern, an arc of shrapnel that terrorized the Mongol horsemen.

Some form of Chinese gunpowder apparently reached Arab nations (probably via traders) during the early twelfth century, for in 1118 some Muslim armies began using what they called a *madfaa*, consisting of a large wooden pot, atop which a stone ball rested. When the gunpowder in the pot was lit, the resulting explosion would launch the ball. Looking something like an egg in an egg cup, the *madfaa* encapsulated every scientific and technical mistake in the book. Wood is the worst possible material to contain explosive gases (as many Muslim gun crews learned, fatally), and the idea of resting the ball atop the firing chamber drastically reduced the power of the exploding gases underneath it. Consequently, the *madfaa* could fire its ammunition only for a short range; almost always, the *madfaa* crews had no idea whatsoever where their shot would land. Arab science was not enlisted to fix the problems, and as a result, the Muslims missed out on the great gunpowder revolution, a lapse that ultimately would cost them an empire.

Meanwhile, however, other civilizations forged ahead. In about 1288, the Chinese made another technological leap, this time learning that if the explosive gases produced by the ignition of gunpowder could be confined in a fireproof and blastproof chamber, the result would be a very strong propellant force, which could be directed outward through an opening in that chamber. Accordingly, they designed an odd-looking iron contraption looking like a large, squat vase, that contained a chamber at the bottom into which gunpowder was poured. A sheaf of arrows was placed at the mouth of the device; when the gunpowder was ignited, the arrows would shoot out at very high velocity.

It was during the campaign to defeat the Mongols when Europe first learned of the new machine of war. A papal ambassador had been sent to the Mongol capital, and his early reports mentioned China's "weapons of fire." Other travelers followed, and soon samples of Chinese rockets, firecrackers, and gunpowder reached the West. Despite Roger Bacon's efforts to keep the gunpowder formula secret, other Europeans gradually deduced it, and by 1300, experiments were under way in Cologne on making a European version of the Chinese weapon. The German experimenters realized, however, that a war engine that could shoot arrows was not quite what the Europeans needed. The real need was for a weapon that could decimate large formations of infantry and cavalry and knock down all those big castles. By 1324, German metallurgists in Metz had made a critical technological breakthrough: a "fire weapon" that could fire spherical stone balls at high velocity. Their design used an enlarged version of the Chinese bottle design, consisting of several sections of short iron tubes joined together by steel hoops—an idea borrowed from barrelmakers.

Initially, these new weapons, called "culverins" and "bombards," fired only round stone balls. Although wildly inaccurate, the explosion, fire, and smoke they emitted—a phenomenon never before seen in a battlefield weapon—were intimidating to any infantry or cavalry formation that first encountered them. The new weapons did not become deadly antipersonnel weapons until 1450, at Montigny, when the French discovered that firing bunches of stones would slaughter mass infantry formations, which sidestepped the weapons' chief fault, lack of accuracy.

But as "wall-crackers," the early guns were something less than intimidating. The problem was that the stone balls tended to shatter against thick castle walls. Armorers in various kingdoms then decided that the answer was larger culverins and bombards, setting off an arms race to develop ever larger guns. The race culminated in a huge bombard that could fire a 600-pound round stone several hundred yards. Nevertheless, that still didn't solve the problem because increasing size failed to address the technological deficiencies that made the bark of these weapons much more severe than their bite. They were inaccurate, highly volatile (an improper mix of gunpowder would result in an explosion that wiped out the weapon's crew), very expensive, and unwieldy. This latter deficit was especially critical, since the lack of mobility meant the weapon could not be effectively deployed on the battlefield. A large gun could require weeks to set up in the field, a task often accomplished by teams of engineers and armorers who assembled it on the spot. It was barely effective for sieges, since most castle walls would not succumb to even the heaviest stone balls launched at them.

The perceived need, then, was for a gun of sufficient range to stand off some distance from a castle (and out of range of any defensive weapons), and with very high explosive power launch ammunition of sufficient velocity to hit castle walls with great force. Also needed was some kind of ammunition that would actually crack apart those walls. As an antipersonnel weapon, the need was for some sort of mobility so the guns could be moved to a new position as changes in the tactical situation demanded.

All those problems were solved by Charles's scientific-technical research teams, although they could not have anticipated the extraordinary changes their accomplishment would bring.

•   •   •

The French effort to develop an ultimate weapon was first spurred by the necessity to checkmate the longbow. It got another major impetus in 1465, when a nasty war between France and Burgundy broke out. The French had a modest technical edge over the Burgundians, and they were determined to widen it, an effort dramatically aided by a conceptual breakthrough achieved by Charles's scientists. They had carefully studied the ever-increasing size of the early guns, concentrating their attention on the ratio between power and velocity, and emerged with a counterintuitive conclusion: smaller was better. It was a dead-end technologically, they concluded, to build bigger and bigger bombards and culverins, for there was no way scientifically that such a design would ever develop the power sufficient to break a thick wall. Besides, the bigger the gun, the more unwieldy it was, not very useful on the battlefield. A giant bombard or culverin, after lengthy preparation just to set it up, might fire one stone ball, followed by equally lengthy preparation to get the thing to fire one more (often it couldn't). Given all the time and expense the big guns required, they weren't very cost-effective.

A more serious drawback to the giant guns, the scientists concluded, was that they were short-barreled, which meant that there was an insufficient area to allow for the expansion of the explosive gases generated by the detonation of the gunpowder. Their solution was a radically new approach: a smaller weapon with a longer "barrel" (a term borrowed from the barrelmakers, whose stave and hoop design was adopted for the first big culverins), a thickened area around the firing chamber to allow for more powerful gunpowder, and tapering the thickness of a longer barrel toward the cannon mouth in proportion to the drop-off of pressure behind the projectile. And that projectile, the scientists concluded, should be a solid iron ball of spherical shape, manufactured to precise specifications that would make it slightly smaller than the barrel to allow for a slight escape of the exploding gases as a safety measure. The gunpowder in the firing chamber would be tamped down tight to provide maximum efficiency for the explosion, followed by the projectile, which would be pushed tight against the gunpowder to provide maximum push when the gunpowder was set off.

But how much gunpowder should be put in the gun? It was a critical technical question, for too much would result in an overpressure that would blow up the gun, while too small an amount would result in insufficient push for the projectile. Mathematicians worked it all out, a set of formulas that specified the precise weight of powder to propel an iron projectile to a given range.

The weapons system was now taking shape, aided by still further scientific and technical breakthroughs. Among the most important, developed by Czech alchemists, was granulated gunpowder, which exploded with much greater force than conventional gunpowder, whose consistency was something like fine dust, one of the factors that tended to make it unstable and subject to temperature extremes. Granulated powder was much safer and functioned in just about any kind of weather. Another technical breakthrough, which ironically enough took its inspiration from church bells, was advanced bronze casting for the barrels of the new weapon. The idea was borrowed from the bellmakers who had developed, after many years of trial and error, very strong bronze alloys for the great bronze bells that rang in church steeples all over Europe. That fact inspired someone (or several people) to wonder: If the bronze in those bells was hardy enough to stand up to the constant stress of being banged by heavy metal clappers, wouldn't it stand up as well to the stress of firing projectiles? It did indeed, although the kind of casting needed to make sufficiently strong bronze was very expensive, which inspired still another research program to find a cheaper alternative.

That effort resulted in a new form of iron that combined the advantages of cast iron (hardness) and wrought iron (workability) by alloying them under high temperatures in new blast furnaces fanned by large bellows. The result, a very hard iron that resisted high temperatures and repeated gunpowder explosions, could be cast relatively cheaply, which meant France could make a lot of the new fire-projectile weapons.

Finally, this entire research and development effort resulted in a weapons system that the French christened "cannon" to distinguish it from its cruder predecessors. With a barrel 8 feet long, the cannon could fire spherical iron projectiles, called "cannonballs," with tremendous force. Trained crews could fire several shots each minute, absolute death on infantry and cavalry. Even better, that kind of power meant there wasn't a castle wall standing that could resist continuous pounding from 50- or 70-pound iron balls smashing into it with tremendous velocity.

The next step in development would convert the new cannon into an even more decisive weapon. The idea was to make the weapon mobile, but that immediately ran into a problem of physics. Isaac Newton hadn't been born yet, so the French scientists had no insight into the scientific principle that every action has an equal but opposite reaction. They did realize that when the new cannons were fired, the explosion created a "kick," meaning the guns recoiled violently backward. Nobody understood why that was so, but the point was that if the cannon were to be made mobile, that phenomenon would have to be taken into account.

The solution was wheels: mounting the cannon on a two-wheeled carriage pulled by horses. That not only meant the cannons could deploy simultaneously with an army in the field, but they also would be mobile (which is why the French called their new cannons "field artillery"). Moreover, the wheels allowed the cannons to rock back in recoil; gunners simply pushed the gun back into position for the next shot when the recoil ended. Finally, another key innovation made the cannons especially effective weapons—a simple piece of technology known as a trunion, a screw-type mechanism used to elevate or lower a cannon to achieve desired ranges. Gunners aimed their guns using firing tables prepared by mathematicians, who worked out all the various angles at which the gun should be placed to attain the necessary trajectories and ranges. Gunners consulted the tables to determine how many turns on the trunion would result in a particular trajectory. (For high-angle fire the French developed a stubby version of the cannon called a "mortar," a wide-mouthed weapon that fired a heavyweight cannonball at short range in a high arc, useful for launching cannonballs over high walls or other obstacles.)

All these innovations resulted in a weapons system that was nothing short of revolutionary, for it not only made every castle and fortress instantly obsolete, but also promised total dominance of the battlefield. Thus it is no wonder that in 1494, when Charles VIII, the inheritor of the French throne and the new technology, crossed the border into Italy, he rode haughtily at the head of his army and its horse-drawn cannons in total confidence that nothing could defeat him.

In the end he wasn't defeated, exactly, but the great wonder weapon he was sure would bring all Italy into his hands was frustrated by that most ordinary of all substances: dirt.

• • •

There exist no records to indicate which Italian mind (or several Italian minds) arrived at the bright idea that the way to defeat cannonballs is to defang them. It was a matter of simple physics: a fortress wall was an immobile object whose strength rested on its ability to absorb punishment. There was a limit, however, to that punishment. A man with a hammer might pound at a thick castle wall for an eternity and achieve nothing except a lot of stone chips. The wall would remain standing because its inherent force could not be overcome by the hammer's much lesser force. However, a cannonball fired with high velocity would hit that same wall and cause it to tremble, because the missile's force was greater than the wall's force. Fire enough cannonballs, and the force of the wall, essentially a pile of stones, will eventually surrender to the greater force.

So the answer lay in finding something whose inherent force would always be greater than the biggest cannonball. No one knows if the answer was arrived at by experimentation or sheer intuition, but sometime in 1500, as Charles was rampaging through Italy, hundreds of workers were busy at fortresses throughout the country excavating huge ditches and piling the dirt behind the outer walls of fortresses. The French invaders were not impressed—until they realized that their cannon were having little effect on these dirt-buttressed walls. They could not have been more shocked: in just six years, their wonder weapon, which had taken nearly a century to invent and refine, wound up being checkmated by men with shovels.

The Italian dirt defense was just the beginning of a long series of frustrations for the French, who finally abandoned Italy in 1529 after years of inconclusive fighting that drew in Spain. But the war left behind a military and scientific revolution.

It took the form, first, of a frantic effort in Italy to invent a new form of fortification that would defeat cannons. Their minds set right by their near death at the hands of cannons, city-states hired the best scientific and engineering talent for the task, which was to result in an entirely new approach, known as *trace Italienne*. It built on the idea of a lot of dirt as a defense: a symmetrical, polygonal fortress with low walls that were sloped (somebody figured out that a sloped object better resists any force moving against it, the idea behind the modern concept of sloped design to overcome wind resistance). Those low, thick walls of compacted earth and masonry squatted behind huge ditches and water obstacles. However, the real strength of the *trace Italienne* was in its geometric design, which allowed defenders to station cannons at key points to provide enfilading fire against every possible angle of attack. The various open areas within the points of the polygon were in fact killing grounds, areas where besiegers would be destroyed in a crossfire.

The new fortifications owed their effectiveness to the science, particularly mathematics, that was necessary to conceive and build them. The leading mathematical achievement was trigonometry, the use of ratios of the sides of right-sided triangles to calculate lengths and angles in geometrical figures, a major tool in such fields as surveying, construction, and cartography. First developed by the Greek astronomer Hipparchus in 140 B.C.E., trigonometry got a major stimulus from the *trace Italienne* system because of the necessity for mathematical precision. By 1551 the mathematical work on the *trace Italienne* enabled the German mathematician Georg von Lauchen to produce detailed trigonometric tables, another essential tool for the scientific revolution.

The *trace Italienne* system had no sooner been built when there were attempts to defeat it (such as a new invention: explosive mines that were laid by a new breed of military technicians called "sappers," who tunneled under the defensive walls and often encountered highly trained "counterminers," who tunneled toward sapper tunnels and engaged the sappers in an underground war of moles). This struggle between attack and defense was only one tremor in an earthquake of military innovation and development that now convulsed Europe. Its cutting edge was science, enlisted to conceive new kinds of strongholds to protect cities, more efficient cannons, more powerful ammunition (including hollow cannonballs filled with explosive charges), new mining technology to fulfill the huge demand for metals, advanced metallurgical techniques, and better tactical systems to take full advantage of the new cannon weapon. The tremors spread deep into the fabric of society: a centralization of power in larger kingdoms with the taxing power necessary to support expensive weapons systems; establishment of permanent military establishments run by military experts; creation of standing armies to keep the essential technicians and engineers on call; and the widening of war to involve entire populations (including the first military draft in 1506, subjecting all males in Tuscany between ages eighteen and thirty to obligatory military service).

The deeper tremor was difficult to spot at first, but it would soon manifest itself in the scientific and intellectual revolutions that would change the world forever. The feverish race to first match the French wonder weapon and then keep a technological edge by constant research and development ignited a spirit of inquiry that was the essential groundwork for the renaissance in thought to come. In an atmosphere where men were encouraged (and paid lavishly) to think without boundaries about superior ways of killing their fellow human beings, that same atmosphere gave them the freedom to think about other things—such as how the natural world really worked. This is not to say that the great explosion in scientific thought that detonated in the seventeenth century stemmed exclusively from Europe's preoccupation with military matters during that time, but it is clear that this great arms race was at least in large measure responsible for it. The advent of gunpowder, cannons, and other revolutionary military technologies had been brought into being by science, and it was science that was essential to maintain and improve it. The military revolution needed new ways to count and calculate, to measure and design with complete accuracy, to observe and record. Such needs gave birth to logarithms, analytical geometry, calculus, statistical methods, navigational aids, and astronomical instruments.

The simple fact is that aside from military considerations, Europe had very little interest, or use, for theoretical science of any kind. It was still an age of superstition, obsessed with the idea of demons lurking everywhere, especially witches. (In 1596, the archbishop of Trier ordered 120 people burned at the stake because, he charged, they had made the cold weather that year "last devilishly long.") At least 80 percent of all Europeans were illiterate, convinced that witches created all the world's troubles—which is why there were more than a hundred thousand executions for witchcraft in Germany alone during the seventeenth century. With the exception of an enlightened few, rulers could have cared less about any kind of science that didn't help them acquire more power or make them militarily invulnerable. They preferred to understand the world through legions of palmists, dream interpreters, and astrologers. Very few kings and princes had science advisers, but almost all had official astrologers to determine celestially propitious times to make important decisions. So did popes. Professors at the new universities that began to spring up in Europe learned to keep their royal patrons happy by issuing annual *indicia*, predictions of future events based on astrology. (When Lorenzo de Medici established the University of Pisa, students demanded a course on astrology.) The best-selling book in 1534 was *Pronostico (Prognostication)*, an immensely popular guide written in vernacular Italian on how readers could predict their futures by using astrology. The author was Girolamo Cardano, an Italian astronomer and contemporary of Copernicus. Cardano had discovered he was about to starve to death working as an astronomer, since no one evinced any interest in such matters as sunspots. To support himself, he went into the astrology business, claiming that his astrological predictions were "scientific." Perhaps, although it should be noted that he predicted that Holy Roman Emperor Charles V would soon die, a prognostication Charles ignored to live another thirty years.

In addition to soothsayers, many kings also had alchemists sitting around in dank corners of their castles, believing that these early scientists, embarked on a futile attempt to find what they called the "philosopher's stone," the secret elixir of all matter, just might be able to transmute lead into gold—which is why they were royally subsidized to spend years in their secret laboratories, trying every conceivable formula in the hope that one of them would magically produce gold. They would have remained among the circle of pseudoscientific crackpots in the king's court were it not for the introduction of gunpowder into Europe. The alchemists, who had spent many years analyzing and mixing chemicals, were in a perfect position to fulfill a new need of their royal patrons: make this new powder easier to handle and more powerful. Swept up in the great arms race inspired by cannons, the alchemists turned their attention to applied science and were able to deduce that black powder, the crude form of gunpowder brought from China, was inefficient as an explosive, since only 44 percent by weight was converted into propellant gases. The rest was solid residue, limiting muzzle velocities to approximately two thousand feet per second, a not very efficient propellant. The alchemists began to experiment (there were more than a few laboratories that disappeared in violent explosions in the process) and emerged with gunpowder mixtures that were much more powerful. To make them safer, the alchemists invented a process that combined gunpowder ingredients in water and ground together as a slurry. Dried in sheets or cake form, the explosive was easier to handle and much more stable than the powder form.

The switch from attempting the transmutation of gold to the more practical task of analyzing gunpowder created modern chemistry, for in the process the alchemists developed the two essential building blocks of chemical science: analysis of chemical properties and controlled experimentation. Thanks to gunpowder, they had become scientists.

In much the same way, the "natural philosophers" of the time—the men we now call physicists—whose solitary musings on such theoretical and hopelessly impractical matters as the motion of the planets had attracted no public or official attention, suddenly found themselves thrust into the world of applied military science. As a result, they would create modern physics. It was not an entirely comfortable evolution for some of them, as the careers of two brilliant natural philosophers demonstrated. Each, in his own way, attempted to reconcile his contribution to war with the notion of "pure" science. Neither man entirely succeeded, an ominous portent of even more profound dilemmas to come.

• • •

Niccolò Fontana has passed into history under the name Tartaglia, in fact his Italian nickname ("stammerer"), which stemmed from a serious speech impediment. Tartaglia had difficulty making himself understood when he spoke, but there was no doubt of his brilliance when he wrote out mathematical equations. From an early age it was clear he had an extraordinary ability for mathematics, exemplified by his pioneering work in cubic equations and the mathematics of falling bodies. By 1535, Chairman of Mathematics at the University of Venice, he was ranked among the leading mathematicians of the age, and for good measure dabbled in various problems of physics. Tartaglia, who abhorred war, considered himself a "pure" scientist and kept himself aloof from the wars that raged around him during those unsettled times. He spurned several lucrative offers to apply his genius in the military service of several kingdoms, but in 1536 he faced a dilemma: Should he use his scientific talents in the service of war to save Venice, a city he loved?

Venice urgently sought Tartaglia's help because it had come under threat from the Ottoman Turks, a threat that was reaching crisis proportions because Venetian soldiers were encountering a very knotty problem. The problem had to do with the new cannons that Venice, like every other political entity on the Continent, had hurriedly built to keep pace with the revolution in armament. The cannons seemed perfectly fine technically, but the gunners were having a lot of trouble hitting targets accurately. Even the firing tables developed by French mathematicians were of little help; consistently, shots fell short or were otherwise wildly off target. Perhaps, the gunners speculated, there was something wrong with the gunpowder or some sort of technical defect in the cannons. Tartaglia thought otherwise. He had never fired a gun in his life (as a dedicated pacifist, he could hardly stand the sight of them); nevertheless, he spent day after day carefully observing Venetian gunners firing test shots from their cannons, which convinced him that there was nothing wrong with the guns. The problem, he realized, ran deeper than that.

It was while watching the cannoneers calculate the range to their targets that Tartaglia grasped where the problem arose: the fundamental law that governed the movement of all objects. The law, which had been articulated by Aristotle nearly two thousand years before, said that all earthly movement is rectilinear: an object thrown (or shot from a cannon) moves in a rigidly straight line, then finally runs out of "impetus" and falls to the ground. This immutable law was the foundation of all medieval science.

However, Tartaglia realized that the great Aristotle was wrong. From his observations, he realized that any object moving through the air—an arrow, a thrown rock, a cannonball—doesn't move in a straight line; in fact, it moves in a curved line, a geometric form mathematicians call a parabola. The curved path is the result of the steady gravitational pull on any object moving through the air, and although Tartaglia didn't know about gravity, he did know that cannonballs irrefutably followed a parabolic path. Cannoneers, adhering to Aristotle's law, had been assuming that range was defined as the maximum point an object, moving in a perfectly rectilinear way, suddenly ran out of "impetus" and dropped, perpendicularly, to earth. No wonder the Venetian gunners—and everybody else—were having trouble with accuracy.

Not content with his observations, Tartaglia worked out the mathematics to prove them. In the process he created an entirely new field of science, the mathematics of ballistics, among science's greatest triumphs. The military implications, however, were momentous: not only had Tartaglia shown how cannonballs actually move through the air, but also the mathematics he devised would now allow gunners to achieve improved accuracy because they could plot, with mathematical precision, the fall of their shots. For those Venetian cannoneers who were mathematically illiterate, Tartaglia devised a gunner's quadrant, an instrument that instantly told a gunner at which angle to elevate a gun to achieve a particular range.

In the supremely tragic irony of his life, the work of Tartaglia the pacifist would unleash the full destructive power of the gunpowder revolution. Aware of the implications of what he had done—and further aware that so useful a military tool inevitably would find its way out of Venice—he seriously considered the unthinkable for a scientist: he would not publish his discoveries. "It was a thing blameworthy," he wrote a friend in anguished guilt about his work in ballistics, "shameful and barbarous, worthy of severe punishment before God and man, to wish to bring to perfection an art damageable to one's neighbor and destructive to the human race."

Finally, however, Tartaglia did decide to publish his discoveries. "The wolf was preparing to set upon our flock," he said, rationalizing that the danger posed by the Ottoman Turks to his homeland justified using his "pure" science in the service of death for a higher cause. Something along that line of reasoning marked the even more glittering career of Tartaglia's scientific inheritor, the man who would be credited with almost single-handedly creating modern science.

His name was Galileo Galilei.

• • •

As did Tartaglia, Galileo had a passion for "pure" science, spurred in part by his reading Tartaglia's work, along with the work of the scientist who would become his greatest inspiration, Archimedes. Galileo felt a certain affinity with the scientific genius of ancient Greece because the Italy into which he had been born in 1564 was, like the Greece of Archimedes' time, constantly embroiled in war; Italian city-states were either fighting with each other or with rapacious neighbors. And like Archimedes, Galileo believed he could have it both ways—serve the military needs of the state while at the same time pursuing his real passion, theoretical science.

In Galileo's case, there were important financial considerations at work. A professor of mathematics at the University of Padua (the only scientific title Galileo ever held) could expect a salary only slightly above subsistence level, and since there was no interest by any patron in funding theoretical science research without practical value, Galileo felt he had no choice: to underwrite the cost of his researches into theoretical science, he would have to offer his scientific abilities to the art of war. In 1597 he set up a thriving business teaching "military architecture" (for which read fortification) to private students, and followed that with a real moneymaker: an instrument he called a "geometric and military compass." The instrument, an early version of the calculator, consisted of a pair of metal rulers, joined by a pivot, containing numbers and scales, along with an attachable arch. Using the arms, almost any angle could be determined, along with extraction of square roots, a handy tool for cannoneers and military engineers, who snatched it up. Its inventor made a killing.

Galileo wrote an instruction booklet (the world's first such document) for the instrument, which he fawningly dedicated to the Medici prince Cosimo, for the simple reason that the Medicis were where the money was. The gesture paid off, for Galileo picked up some extra cash when the Medicis hired him as a private tutor for Duke Fernando, future heir to the Medici throne. (The job abruptly ended when Galileo was summoned, to his distress, to cast a horoscope for Fernando when the duke fell ill. Galileo politely foretold "many more happy years" for his pupil, who died three weeks later.)

Galileo also filled his coffers by contributing his talents to the

Venice arsenal, at the time the world's most advanced naval research center. The Venetians, who ran a lucrative trading empire, were determined to protect their precious merchant fleet with state-of-the-art naval technology. That determination took form in the arsenal, lavishly underwritten by state funds, which recruited the best shipwrights, naval architects, and scientists. The brightest star in this glittering array of scientific and technical talent was Galileo, who among other things came up with the idea of testing new plans for ship hulls and warship designs with scale models, along with the new procedure of systematically extrapolating data from tests using the models. This scientific approach to naval design came to be the prototype for all military research and development laboratories since then—an idea, theoretical work on paper to determine if the idea might work, production of a scale model to test the basic concept, and finally a full-scale prototype subjected to rigorous testing before full production.

Galileo didn't miss a trick. In 1609, when he heard about an amazing new Dutch instrument called a telescope, he promptly ordered one, built an improved version, then lugged it to the top of the highest building in Venice. There, he brought the doddering old *doges* who ruled Venice to gaze through the instrument—which, Galileo pointedly noted, would allow them to "discover the ships and sails of the enemy two hours before he can become aware of ours." The delighted *doges* awarded him 1,000 gulden (about $13,000) on the spot. That very night, a financially flush Galileo turned that telescope toward the heavens and began the observations that would make him a science immortal. The money also allowed this remarkable virtuoso to conduct the experiments on motion that founded modern physics.

Significantly, there is no mention of any military work in Galileo's writings, nor does he mention that near the end of his life he attempted to interest, unsuccessfully, major naval powers in the idea of using his observations of the satellites of the planet Jupiter as a navigational device. The science writings that got him in hot water with the Inquisition (not, as commonly believed, for arguing that the earth revolved around the sun, but because he couldn't resist taking a dig at Pope Urban) all concern his first love, theoretical science.

History would put Galileo in the pantheon of human thought, but in his world he was simply another scientific resource to be exploited for whatever he could contribute to applied science, the essential tool that kept political entities alive. Europe had fallen into a mad orgy of religious wars, civil wars, economic wars, and political wars, an endless cycle of conflict that created an insatiable need for better weapons, the precious technological edge. Mines were working nearly around the clock to produce more and more ore for the growing war machines, while every kingdom and city-state that could afford the cost harnessed science to the task of producing better ways for men to kill each other. In this cycle, the fortunes of kingdoms fell and rose in proportion to whatever technological edge they might devise, demonstrated by the sudden rise of Spain to the status of a major world power. That happened because some bright Spanish scientists got the idea that the best way to defeat the wonder weapon of the cannon was simple: kill the cannoneers.

• • •

Spain, which had its own imperial ambitions, looked with alarm on the miraculous cannons its chief rival, France, had brought to Italy. The Spaniards immediately began recruiting topflight scientific and technical talent (including Tartaglia, who said no) to come up with an antidote. By 1501, several Flemish armorers who had gotten into the new field of ballistics science arrived at a critical insight: the weakness of the cannon was not in the weapons themselves, but in the men who fired them. Standing 100 yards or more away from an infantry formation, the cannoneers were not vulnerable to short-range infantry weapons (except the longbow), but suppose the infantry were equipped with a weapon that could endanger the cannoneers?

The result was what the Spaniards called "fire stick" or "hand cannon." Fundamentally, it consisted of a minicannon that would put the range and firepower of a cannon in an ordinary soldier's hands. The first examples of this new weapon were crude, amounting to small-size cannon with a 20-millimeter bore and weighing up to 20 pounds that required a forked rest to use. It fired a 2-ounce ball at a range of up to 100 yards, a process that seemed to take forever to complete (some ninety separate steps were needed for the wielder of the weapon to load, aim, and fire it). However, the Spaniards made rapid technical improvements, including a matchlock—a mechanism attached to the side of the weapon that touched a burning match to the powder by means of a new invention called a "trigger"—and a wooden bent stock to allow an infantryman to shoot the weapon from his shoulder. Called an *arquebus* ("hooked gun") by the French, this device began to transform the "hand cannon" into a powerful infantry weapon. Within two decades, the early "hand cannon" had been developed into a terror weapon against artillerymen and became the foundation for the next revolution in military science—an elongated and redesigned model called a "musket." "Musketeers," as the men who fired these weapons were known, could get off only two shots every three minutes because of the lengthy process of putting powder in the firing chamber, tamping it, inserting a musket ball, priming the firing mechanism, aiming, and then firing. But the Spaniards organized these musketeers into highly trained infantry units whose ranks could alternately load or fire, creating steady volleys of fire fairly accurate up to 300 yards—fatal to artillery crews who got anywhere near them.

Just how deadly the new weapon could be was demonstrated on February 25, 1525, when a force of fifteen hundred Spaniards trained in the use of the new "hand cannon" were attached to Spain's ally, the Hapsburg emperor Charles V, for an attack on the French army in Italy. The French and their cannons were in the process of reducing the fortress at Pavia, in northern Lombardy, to rubble when Charles's army showed up to drive them away. The French attacked and immediately discovered that their vaunted cannoneers were being cut down before they could deploy anywhere near the Hapsburg infantry. The French first assumed they had encountered new cannons the Spaniards somehow had developed in record time, but finally realized they were up against something odd: troops firing what appeared to be small cannons resting on forked sticks. The French cavalry was sent in to sweep them off the battlefield. But the disciplined Spanish gunners, trained for just this eventuality, stood their ground and shot the cavalrymen from their mounts before they could get within 50 yards. By the end of the day more than eight thousand French soldiers had fallen.

News of the defeat sent a new shock wave through Europe: still another technological leap forward, still another wonder weapon to contend with, still another effort to keep pace. Even worse, this was getting expensive: a matchlock firing mechanism for the new guns cost $600 in sixteenth-century money, and there wasn't a king who couldn't figure out how much of his treasury would be required to keep up with even this small piece of new technology. And then there was the cost of all those people needed to keep the edge sharp—the craftsmen who built the new weapons, the gunsmiths who kept them in working order, the scientists working to develop new weapons, the technicians perfecting new firing mechanisms, the chemists devising improved gunpowder.

However ruinous the cost, kings would come up with the money, for if there was one thing they had learned since Agincourt in 1415, it was the supremacy of the scientific cutting edge to maintain and extend their power. No king could run the risk of losing his army in an afternoon to a rival whose army was equipped with a new wonder weapon.

This was the impetus for the steadily accelerating military revolution in Europe, the twin brother of the scientific revolution that swept through the Continent at the same time. It was a leapfrog process, however, that did not take root to the east, although the great powers of Islam and China were just as interested as their European counterparts in power and conquest. But unlike the Europeans, science in the East had entered a period of stagnation. And that would prove fatal to their very existence.

As the Europeans forged ahead scientifically and militarily, Islam had no insight into the rapid pace of scientific and technical change those two intertwined revolutions implied. In 1453, the Islamic armies had taken Constantinople for good, thanks in part to their use of huge *bombards* that had shattered the city's massive fortifications. Very impressive, but the fact was that the big guns were not products of Islamic science; they had been built by Hungarian technicians first recruited by France to help develop cannons, and then subsequently hired by Islamic armies to solve the problem of pounding through Constantinople's outer ring of defensive walls. As it had done so often before, Islam borrowed various scientific developments—gunpowder from the Chinese, and now *bombards* from the Europeans—and then rested content, believing that whatever ultimate weapon they were using would remain supreme. There were plenty of qualified Islamic scientists, but their knowledge was tightly restricted inside the ruling classes, where rigid religious orthodoxy prevailed. Those ruling classes regarded the printing press as a dangerous instrument of subversion, especially in a time of deep divisions among the Sunni and Shi'ite religious factions, so any new ideas of Islamic science weren't disseminated. The penalty was paid in the Ottoman War of 1593–1606, when Islam was routed by the gunfire of Europe's advanced science and technology. The Islamic kingdoms began a long period of decline and final collapse.

Religious orthodoxy also constricted any flowering of Chinese science, which was responsible for the invention of virtually every technological advance later associated with the scientific revolution—printing, gunpowder, and metallurgy, to name just a few. But Confucianism was the official ideology, and its stress on ethics, morals, filial piety, and acceptance of the world as it was scorned science as an unholy questioning of eternal moral values. Essentially, Confucianism argued that mankind should adapt itself to the universe, not try to change it.

Additionally, China went through periodic convulsions of xenophobia and insularity, during which Chinese science came to a standstill. In 220 B.C.E., while the Western scientists at Alexandria were making fundamental discoveries about mechanics and how the universe worked, Tsin Jeng, the first empress of China, was ordering the burning of all books as dangerous temptations to people who would question eternal wisdom. When appalled scholars protested, the empress had five hundred of them buried alive, an event that makes Galileo's later troubles with the Inquisition seem positively benign by comparison. The use of movable type for printing (using wooden letters instead of the cast metal system later devised by Gutenberg) had been first developed in China in 869 C.E., but books were circulated only among a small group of scholars, and type was destroyed after printing to prevent it from being used to print dangerous ideas.

In China, truth had been made manifest, and there was no reason for the "Kingdom of Heaven" to inquire any further; and as one Chinese scholar announced, "No more writing is needed." Secure behind their Great Wall, the Chinese believed they alone were civilized, and that they could keep out the "barbarians" while retaining the secrets of their science. But given the gradual development of trade between China and the outside world, neither of those goals was attainable. It was just a matter of time before the barbarians learned of such things as the secret of making silk and the formula for gunpowder. The Chinese persisted in believing that science was some sort of sideline hobby of the ruling entity that could be stored away in a closet until needed again, and in the process kept locked away and forever secret. However, among other things, they forgot about economics: If a Chinese trader was offered a lucrative deal to provide some samples of Chinese military technology, it can be safely assumed that technology would find its way out of China.

By the sixteenth century, the Chinese at last began to realize that centuries of xenophobia had put them far behind the European barbarians scientifically and technologically—and it was too late for China to catch up. In 1578, a Jesuit named Matteo Ricci, representing the Vatican, arrived in the Forbidden City of Beijing to begin work as a missionary. To smooth the way among the insular Chinese officials of the Ming Dynasty, he brought along several gifts, including one that flabbergasted its recipients. It was an early triumph of European technology, the spring-driven clock. "A clock that rings of itself!" one awestruck Chinese official remarked, with more than a hint of nervousness: if these Europeans were such ignorant barbarians, then how had they managed to devise something so wondrous, a marvel that even Chinese science had not been able to produce?

Chinese nervousness increased over the next several decades as growing numbers of European traders and missionaries arrived with even greater wonders, such as cannons cast of bronze alloys (how, the Chinese wondered, could weaker minds than their own invent a metal that never rusted?). Finally, on the bright, sunny day of June 21, 1629, the Chinese were given a graphic lesson in just how far they had fallen behind the barbarians.

Several Jesuit missionaries, who had received scientific training, informed the Chinese emperor that he might want to go outside that day to witness a solar eclipse that would occur precisely at 11:30 A.M. and last exactly two minutes. The emperor consulted his court astronomers: Was it possible that these barbarians were now able to predict celestial events, a science that, as was well known, only Chinese genius was capable of mastering? The court astronomers smiled tolerantly: no, the Europeans could not possibly know of such things. In fact, they announced confidently, there would be a solar eclipse that day, but it would take place at 10:30 A.M. and last two hours.

Just before that time, the emperor and his court, dressed in the elaborate robes of office, gathered expectantly in the great courtyard of the Forbidden City and looked skyward as the court astronomers counted off the minutes. At ten-thirty, the sun shone brightly—and continued to shine for the next hour. The emperor and his retinue, by then increasingly agitated, waited. At precisely eleven-thirty, the Forbidden City was plunged into darkness as the solar eclipse began, just as the Jesuits said it would. And exactly two minutes later, the sun shone again.

"You are the sons of heaven," the emperor proclaimed with a slight bow of respect to the Jesuits. He then brushed by his mortified court astronomers and adjourned to his throne room, there to sit morosely among his unsettled advisers and consider the consequences of what he had just witnessed. No one knew how these barbarians had managed to achieve the impossible feat of surpassing the science of the Mandate of Heaven, but one thing was clear: European science was supreme. And it was equally clear that China was far behind; what the consequences of that would be, none of them dared guess.

Actually, the tide of history now running against them was more ominous than the Chinese knew. What they did not yet understand was that the Europeans, in the throes of a scientific and military revolution whose costs were astronomical, were moving aggressively outward for the treasure they needed to feed that roaring fire.

4

Outward Bound

Gold is a wonderful thing. Whoever owns it is lord of all he wants. With gold it is even possible to open for souls a way into paradise.

—CHRISTOPHER COLUMBUS

The mighty army of eighty thousand that had gathered in the Peruvian highlands town of Cajamarca, ready for battle in body armor of woven hemp, brandishing clubs and slingshots, waited impatiently in the predawn darkness for the sun to rise, when their emperor would speak to them. Promptly at dawn on November 16, 1532, Atahuallpa, resplendent in his robes of office, stood to face the sun, its rays glinting on the magnificent sun emblem of gold on his chest. A hush fell over the army as the absolute ruler of millions in the Empire of the Incas—the God of the Sun, the Immortal One, the Supreme of Supremes, the Invincible One, the Holiest of Holies, the Mightiest of the Mighty—began to speak.

The army bowed their heads in respect as their living god told them his divine will had decided the foreign intruders must be destroyed. The army would now move to the plain below and kill the usurpers, the evil ones whose unholy ways so offended the gods. The bodies of these invaders would be ripped apart, their hearts and livers offered as sacrifices to assuage the displeased gods. With a wave of his arm, Atahuallpa sent his army on the attack, and in a mass, they moved toward a small group of Spanish *conquistadores* gathered on the plain below the town.

The Spanish commander, Captain Francisco Pizarro, calmly watched the approach of the enemy host. When they had approached to within several hundred yards, Pizarro deployed his

167 men in a rough semicircle, with a few dozen cavalrymen posted at the wings and 30 *arquebusiers* in the center. All his men were outfitted in the standard sixteenth-century European military style: steel helmets, steel breastplate armor, and double-edged swords.

When the Incan army got within 200 yards, Pizzarro went on the attack. While his cavalry swept into the Incas from the left and right flanks, his *arquebusiers* let loose a volley into the packed mass. Then the rest of the Spaniards rushed forward, slashing and stabbing with their swords. Within minutes, the Incas, frozen in fear, became a panic-stricken mob, too terrified to use their weapons as the Spaniards slaughtered them. The Incas were certain they were being assaulted by gods: never having seen horses, the combination of man and mount appeared to them as a double-headed god. Even more terrifying were the guns; these men must be gods, because only the gods had thunder.

When more than eight thousand of them had been killed, the rest of the Incas fled in all directions. The Spanish didn't suffer a single casualty. Atahuallpa, the Invincible One, had watched the slaughter in anguish, now convinced these must be gods who had decided to punish his people. When Pizzarro's men seized him, he was shaking in terror. Prostrating himself before the conquerors, Atahuallpa listened uncomprehendingly as Pizzarro announced, in what he meant as a macabre joke, "We Europeans are here because we suffer from a terrible disease for which there is only one cure: gold!"

Eventually the Incan emperor would come to understand what Pizzarro meant. The Spaniards held him ransom, demanding that the Incas bring them every piece of gold in the empire. Convinced that the ransom would somehow appease these angry gods of thunder and double-headed spirits, the Incas meekly complied, eventually collecting enough gold to fill a room 8 feet high, 32 feet long, and 17 feet wide. But the price to the Incas of the Spanish invasion would be much greater: the end of their empire, destruction of their civilization, and enslavement by the *conquistadores* to produce more gold.

The destruction of the Incas by so small a force was due, primarily, to the superior armament of the Spaniards, the fruits of European science: steel swords, steel armor, guns. However, there were deeper reasons at work: the inability of the Incas, first, to understand the grave danger they faced, and second, their failure to adapt their rigid theocracy to the drastically altered circumstances that the Spanish invasion represented. By the time the Spaniards had decided it was time to deal with the Incas, they had been rampaging through Mesoamerica for nearly two decades, crushing one Mesoamerican empire after another. Each was overwhelmed by the superior Spanish science and technology, yet by the time Pizzarro arrived in Peru, the Incas knew nothing about his intention to destroy them, nor did they know anything about his cavalry, steel, and guns. That's because none of the Mesoamerican civilizations had a written language, so there was no way for the Aztecs of Mexico, for example, to convey the full dimension of the threat from the *conquistadores*—no messages, no bulletins, no reports, no analyses, nothing that rang the alarm bell. Small groups of *conquistadores* simply entered the territory of a Mesoamerican civilization (to that civilization's total surprise), demanded all their gold, and when a Mesoamerican army tried to stop them, coolly used superior technology to slaughter forces many thousands of times their size. It appears that none of these civilizations ever considered the idea of carefully analyzing the Spanish threat in an attempt to understand how the weapons of the *conquistadores* could be countered, or whether they should adopt different tactics.

The Spaniards, on the other hand, had a powerful weapon in their written language. It was the instrument by which Christopher Columbus could convey the importance of what he discovered on October 14, 1492, when he landed on a small island in the Bahamas, and it was the instrument that the later wave of Spanish explorers used to convey the scale of riches available in South America. And it was the means by which Spain organized the military expeditions to get those riches, the sailing orders that told them how to get there, and what to expect on arrival.

The Caribs, destroyed by Columbus, were among the first of a number of ancient civilizations in the New World to be overwhelmed by a scientifically superior civilization that had escaped from its bonds and was now embarking on a great voyage to explore and exploit a world they believed was theirs for the taking. They intended nothing less than establishing a domination over new seas and lands—to escape from the small inland sea of the Mediterranean that had preoccupied them for so many centuries, to expand their horizons, to explode out of their small kingdoms, and above all, to make their fortunes.

The flash of guns at Cajamarca that so terrified the Incas illuminated just one part of the tragedy that would be enacted on three continents. Civilizations locked in the confines of theocracies that had convinced their people there was no reason to change or adapt could not even begin to understand the threat from a civilization for whom change and adaptation had become as natural as breathing. Their destruction was ordained in the minds of brutally pragmatic men who knew how to use the tools of conquest— the compass, the astrolabe, the sword, the cannon, the armor, the ship, the book. These triumphant examples of advanced European civilization had not been invented specifically for conquest, but since their roots lay in war, it was inevitable they would be used someday to destroy other, weaker, civilizations that happened to be in the way.

It was an inevitability that was born 117 years before that bloody day at Cajamarca, when a Portuguese aristocrat realized he was standing in the middle of a gold mine.

• • •

In 1415, at roughly the same time that the English longbowmen were killing the feudal order at Agincourt, the twenty-one-year-old son of Portuguese king John I was assaulting the Moorish stronghold of Ceuta, in North Africa just across from Gibraltar. The Moors had been driven out of Spain and Portugal, and now John wanted them out of Ceuta, too; the stronghold represented the constant threat of a base from which Moorish forces might launch future assaults against his kingdom. The prince, Henry of Avis, a bright young aristocrat with a scientific education, had been given the assignment by his father, who wanted to train the heir to the throne in the art of waging the kind of military campaign he would need to master in a time of constant war.

Henry succeeded brilliantly, taking Ceuta in only one day— thanks to a wonder weapon he had brought with him: a small force of English longbowmen lent by the English king (who regarded Portugal as a counterweight to the growing power of Spain). Terrified Moors fled in fear when the longbowmen opened fire at a range of

200 yards, picking off defenders from the stronghold's walls before they could even unlimber their weapons.

Inside the abandoned city, Henry and his troops immediately began plundering and found enough gold, silver, and jewels to fund the Portuguese royal treasury for the next three decades. But there was an even greater treasure in Ceuta, Henry discovered: tons of spices, including cinnamon, pepper, cloves, and ginger, for which Europe had developed an insatiable appetite since the days of the Crusades. This vast store of spices would turn a handsome profit when sold on the European market, but Henry was thinking further ahead—the real profits would come, he realized, to the kingdom that could monopolize the trade routes that brought these spices westward from Asia. At that moment, Henry decided Portugal would be that kingdom.

The decision would have immense military and political significance. It would make Europe the master of the world, begin the age of exploration, revolutionize naval warfare, and immortalize Henry of Avis, the man who started it all, as Prince Henry the Navigator. But tragically, it would also destroy entire populations and enslave millions of others. These consequences were born of science— although it is fair to say that none of the scientists who decided to help Henry become a monopolist could possibly have foreseen them.

• • •

Henry's key insight centered on how he proposed to dominate the spice trade. Ceuta was at the terminal end of lengthy land caravan routes that extended all the way into China, routes so busy that Henry counted nearly twenty-four thousand trading stores in the city that handled all the transactions the booming commercial traffic generated, thanks to the burgeoning European market for spices. But, as he also realized, those trade routes represented a problem: they ran through Muslim-controlled lands, which meant Muslim satrapies could extract various "fees" and "taxes" or, even worse, cut the trade routes off from Europe altogether in the event of war.

So the solution, Henry concluded, was seaborne trade routes to avoid the entire Muslim problem, both on land and in the Mediterranean, ringed by Muslim kingdoms. In his conception, fleets of Portuguese cargo ships laden with the kind of finished goods the

Chinese and other Asian markets wanted—the cloth produced by Flemish looms, Venetian glass, German metal—would sail from Portuguese ports down the western coast of Africa, around the tip of that continent, then into the Indian Ocean trading centers, and return with hulls stuffed with spices.

Henry believed that the spice trade would make Portugal a mighty maritime power, but as he was all too aware, there was much work to be done before fleets of Portuguese ships sailed down the coast of Africa. A number of hurdles had to be overcome first— not the least among them the problem of falling off the edge of the earth.

• • •

Despite the accelerating pace of scientific advances, by 1415 science had contributed little to the considerable task of navigating ships on open seas. For many centuries, ships had depended on some early scientific advances to learn where they were going—the ancient Greeks had utilized geometry to orient navigators attempting to determine how far they were from land, and ancient astronomers had taught navigators to steer by using certain celestial signposts, notably Polaris, a fixed point in the northern sky that always indicated north.

But that still made navigation an uncertain art at best. Steering by the sun was fairly reliable—except on overcast days, or at night. In the dark, Polaris was a handy signpost, all right, but what about a stormy night, when the star could not be seen? Worse, some of the braver Portuguese sea captains at the beginning of the fifteenth century had tried to sail down the western coast of Africa and discovered that at some point, Polaris literally disappeared from the sky—an unsettling event that convinced them that somewhere around the middle of the continent's coast, the world abruptly ended, much like the edge of a table.

What the Portuguese captains didn't realize, of course, was that as they sailed southward down the West African coast, Polaris dropped toward the northern horizon behind them, disappearing altogether when they reached the equator because of the curvature of the earth. All the Portuguese knew was that the disappearance of Polaris made their return trip navigationally impossible, since they did not have Polaris available as a guide (and had no knowledge of any southern equivalent, since early astronomers made their observations only in northern latitudes).

Given these problems, ships tended to sail in pretty much the same way the Phoenicians, Greeks, Carthaginians, and Romans had sailed centuries before: close to land. As long as a ship was within sight of land, a navigator could take landmarks and determine, with relatively good accuracy, just where the ship was. That worked reasonably well in such closed seas as the Mediterranean and the Baltic, or routes that hugged the European coast, but would be of no use once a ship entered the open seas, where there were no landmarks and where celestial navigation was subject to the whims of weather.

No one was more aware of these defects than Prince Henry, who decided they would have to be overcome before Portugal had any hope of achieving maritime greatness. And the solution, he concluded, would be science.

• • •

In the summer of 1420 the people of Sagres, a seaport on Portugal's southwestern coast, were puzzled by some strange goings-on: an entire horde of men lugging trunks of papers moved into a series of buildings under construction on the high cliffs overlooking the harbor. The men were not very enlightening about what they were doing; unusually tight-lipped in the traditionally open world of seafaring, they settled into the buildings, kept strictly to themselves, and seemed to spend most of their time poring over those papers and writing their own papers. In the harbor, it was noticed that a number of strange-looking craft were under construction, types of large ships no one had ever seen before. And there were workshops that seemed to be busy around the clock with the sounds of hammering and sawing.

What was happening at Sagres represented a large-scale scientific research and development effort that rivaled Alexandria. Thanks to the riches seized from the Moors, Henry had the deep pockets from which he paid top dollar to lure a glittering collection of scientific and technical talent—Greek shipwrights, Italian mathematicians, Jewish astronomers, Phoenician ship designers, Muslim cartographers—to a research institute he founded at Sagres with instructions to develop an entirely new design for oceangoing ships and the navigation tools that would allow them to sail anywhere with scientific exactitude. He decreed that the effort was top secret; they were not to breathe a word of what they were doing at Sagres, and they were never to use whatever knowledge they gained there on behalf of any other kingdom.

Over the next forty years, this intense research effort would revolutionize all human existence because the scientists and technicians of Sagres were able to lift the clouds of error and superstition that for many centuries had constricted exploration of the world. The clouds had been created by Aristotle and Ptolemy, who claimed that the sun was so close to the equator that that area of the planet was all boiling seas and desert—and ships would disappear in the cauldron. Sailors had long assumed that the earth was probably round, but were intimidated by what they believed was a huge bulge at its center. This bulge, they were certain, contained seas so thick with salt that no ship could cleave it. And they also believed that because of the great bulge, any human beings living on it were misshapen savages (young seamen were solemnly informed by their older shipmates that any Christian attempting to sail through the bulge would be turned into a Negro).

Such horror stories were regarded as fairy tales by the scientists working at Sagres. They had already concluded that earth was an imperfectly round globe, with the equator marking the exact halfway point in the globe's north–south direction. True, it was probably warmer in equatorial latitudes, but as the astronomers pointed out, the sun did not approach the planet close enough to produce anything like the kind of heat needed to boil seas. They had conducted a number of experiments that revealed the large amount of heat required to bring even a small amount of seawater to a boil (mainly while trying to find out a way to distill salt), and the kind of heat needed to bring fathoms of seawater to a boiling point was beyond calculation. If the sun was hot enough to boil water at the equator, then it logically followed that such intense heat would have turned Europe into a desert. And it hadn't.

Only a true oceangoing ship would prove whether the scientists were right, and that is where the main effort at Sagres was concentrated. The result was a revolutionary ship design that would open the world's oceans to exploration. Called a *caravelle,* it was a 50-ton design that featured decks high above the waterline, multiple masts, a combination of triangle-shaped lateen and square sails, and a deep hull (to carry a lot of cargo). In technological terms, it represented that century's equivalent of the switch from horse-drawn wagons to vehicles powered by the internal combustion engine. For centuries, the ideal ship design was considered the low-hulled galley powered by oars, supplemented by a lateen sail, a design the scientists and technicians at Sagres concluded would never work on the open sea. Galleys were too slow (an important factor when shipping goods); unseaworthy in rough seas (as the ancient wrecks still littering the bottom of the Mediterranean attest); too labor-intensive (it's hard to recruit rowers, unless they're slaves or convicted criminals serving out terms of penal servitude); and, most significantly, technologically unsuitable for the mounting of the new wonder weapons, cannons (a galley's center of gravity could not handle the weight of heavy guns).

The *caravelle* changed all that. It was very sturdy, able to handle even the roughest seas, very fast, could haul a lot of cargo (or store a lot of provisions for long voyages), and had reinforced decks to allow for the addition of guns. But the real genius of the new design lay in its propulsion and navigation systems. Extensive testing had shown that the ideal propulsion for a sailing ship consisted of a combination of the triangle-shaped lateen sail that had powered ancient Mediterranean fleets since the time of Troy, working in tandem with a square sail. This design utilized the advantage of a lateen sail (maneuverability) while adding the advantage of a square sail (the best design to catch the maximum amount of wind). The new propulsion design also eliminated the lateen sail's chief weakness, insufficient power; ships with only lateen sails were prisoners of prevailing winds—not a problem in the predictable wind flows of the Mediterranean, but useless in the more unpredictable winds of the Atlantic and other oceans.

The more striking scientific advance was reflected in the ship's navigation system. First, the Sagres designers devised a sternpost rudder connected to a steering wheel, which made a sailing ship much faster, since a helmsman could now tack to take advantage of shifting winds with just a turn of a wheel. Moreover, he could keep the ship on course using the next Sagres innovation, a wondrous instrument called a compass, a revolutionary piece of technology whose needle always denoted north–south. The basic concept of the compass was borrowed from the Chinese, and meant ships could sail under cloudy skies, guided by an instrument unaffected by weather conditions and time of day or night. And finally, an entirely new innovation: sailing charts, the first attempt to map the seas. The early charts were not especially accurate due to lack of surveying data, but did include at least some of the important information any navigator needed to know: landmarks, compass directions, sea depths, and navigational hazards.

By 1435, Prince Henry's new ships were cautiously poking their way to the equator, where the nervous crews made a happy discovery: lands around the equator weren't deserts, they were lushly green. Six years later, in the process of developing shipping routes farther south along the western coast of Africa, two of these ships opened what would prove to be a dark chapter of history. They anchored off the coast one day and sent armed parties ashore to explore the lush vegetation. They immediately encountered several natives who fell to their knees in terror upon seeing these strange creatures. The Portuguese put them in chains, announced their baptism in the Christian faith, then bundled them aboard the ships as slaves. They were later put to work in Portuguese plantations, where their free labor so dramatically increased profits, Henry was soon bombarded with requests to bring back more—for which, of course, Henry noted, the plantation owners would be expected to pay. The result was the first great profit-making enterprise for Portugal's merchant classes, trafficking in slaves. By 1448, some nine hundred African slaves had been brought to Portugal, just the beginning of what would eventually become a flood of enslaved human beings.

At the same time, another dark chapter was being opened: imperialism. For all his scientific enlightenment, Henry was a man of his time. Like his fellow European rulers, Henry believed fervently in the idea that the Europeans had a superior civilization that was to be imposed upon "heathens" and "savages" who needed to be converted to European religious and political orthodoxy. And if these barbarians happened to occupy land Europe coveted for its burgeoning economic and military interests, then it was perfectly justified for the Europeans to take the land in the name of "civilizing" inferior peoples.

The *caravalle* represented the best instrument to effect that policy, as demonstrated by Columbus, whose *Santa Maria*, a *caravelle* escorted by two smaller ships, conquered the land he christened El Salvador. Although he was not able to achieve his dream of finding vast gold deposits that would justify the royal investment that had propelled him westward, he set the pattern for European imperialism: a landing upon strange shores, a pronouncement that all the land was now within the domain of a particular European kingdom, enslavement or slaughter of the native population, and then exploitation of whatever natural resources could be found there.

But despite the great advantages of the *caravelle*—its ability to sail long distances, stand up to the rigors of lengthy ocean voyages, and carry plenty of cargo—from the moment the prototype sailed out of the harbor of Sagres, the new ship created two major headaches. One was protection. *Caravelles* loaded with valuable cargo would be prime targets in the event of war (they also would attract the attention of pirates). That meant the ships would have to be fitted with some sort of protection, enough firepower to deter predators. The design specifications of the *caravelles* included the necessary structural strength to arm them with cannons, but nobody had yet figured out how to put heavy cannons aboard ships. The problem was that mounting heavy cannons on a merchant ship's deck created serious instability problems, threatening the ship's center of gravity. Moreover, there was the problem of recoil; when fired, a heavy cannon recoiled backward several feet, and there was very little room for recoil on a crowded deck. Consequently, *caravelles* could be outfitted with just a few small cannons, useful only for signaling. The truth was that to be invulnerable, merchant ships would have to serve double duty: merchant ship and man-of-war. One alternative was to build new warships armed with heavy cannons and capable of keeping up with merchant ships while they did escort duty. But warships faced the same technical problem: How to mount heavy cannons on the upper deck?

Some of the best scientific and technical minds in Europe were applied to this problem, but the solution was elusive: no one was able to come up with a way to put the new French cannon designs on ships without destroying their seaworthiness. Even more intractable was the recoil problem: how to handle the cannon's recoil in the narrow confines of a ship's deck. For good measure, there was also the problem of how to make a cannon shoot accurately from a moving ship that might be rolling in the waves.

The solution would finally come, but in the meantime, there were even more immediate problems involving how to ensure that the ships got to where they were supposed to go. The new compass began service with great promise, but as Columbus reported with some alarm on one of his return voyages to Spain, there was a mysterious glitch in these new instruments: their needles often stopped pointing toward Polaris. Since Polaris hadn't moved (or so the astronomers said), that meant only one thing: the compasses didn't work properly. Then there was the problem of the early charts. They were relatively accurate for any water near land, but largely useless on the open sea. What was needed, clearly, was some kind of new chart system that would allow navigators plotting courses on the open sea to calculate from a chart just how far their ships had traveled. (One of the tales seafarers liked to relate about the vagaries of early charts concerned the Portuguese ship that relied on one such chart to sail to North America, only to find itself some 2,000 miles off course, eventually bumping into a large landmass called Brazil, which the ship's captain promptly claimed for Portugal.)

A major step toward solving the chart problem was made by Paolo Toscanelli, an Italian mathematician and cartographer whose family just happened to be in the spice business. Aware of Prince Henry's efforts to monopolize the spice business for Portugal, and further aware that Henry's scientists were feverishly trying to develop a workable new system for sea charts, Toscanelli arrived at what he assumed would be a moneymaking idea: develop that system on his own and sell it to Henry. Toscanelli worked nearly around the clock on the the task, aided by his extensive reading in the ancient geography texts of Ptolemy. Those texts revealed that Ptolemy had developed a crude grid system for some of his early maps; those using the maps could determine how far they had moved by counting off segments marked on the individual squares of the grid. That gave Toscanelli an insight, combining perspective geometry with a similar grid system. He reasoned that if perspective geometry allowed measurement of an object at distance, the same could be done for the surface of the earth. He then prepared a series of maps with land and sea areas overlaid with a series of grids. With these maps, a navigator could move from point to point by simply marking off a given number of grid distances while maintaining direction by means of the sun (always guaranteed to rise in the east and set in the west) or Polaris. A brilliantly simple system, reflected today in the gasoline station road maps that are marked off in grids, with letters at the top and numbers running down the side. An index shows that, say, West Podunk is at G7. Anyone looking for West Podunk simply finds "G" horizontally, then the point "7" vertically; West Podunk is where the letter and the number intersect.

Convinced he had a winner, Toscanelli rushed to the Portuguese Committee for Navigation, as Prince Henry's institute at Sagres was officially known, and presented his new charts. To his shock, however, the committee rejected the idea. Why it did so remains a mystery, but the more important fact is that a miffed Toscanelli then shopped his idea to Portugal's chief rival, Spain, which immediately snapped it up, at a premium price. For Spain, Toscanelli's new chart system arrived just in time, for the Spaniards had decided they must begin a strong effort to match the Portuguese seaborne offensive. (One of Spain's first major efforts was Columbus's voyage, which used Toscanelli's charts. By an odd quirk of history, Columbus accidentally discovered America because those charts were wildly wrong: the calculated distance from Spain to India was about 3,000 miles off, since Toscanelli was unaware of a landmass called the Americas.)

Spain's eagerness to snatch up Toscanelli's new charts represented the first indication that despite Henry's efforts, he could not indefinitely maintain the scientific and technical lead that had made Portugal the first great imperial power. There simply was too much potential treasure out there for other nations to sit idly by, watching the Portuguese rake in the profits. As the voyages of Columbus demonstrated, the Spaniards were embarked on a program to fulfill their own imperial ambitions, and were willing to pay whatever was necessary to achieve it. The Spanish began a research and development program modeled on Prince Henry's Sagres institute, an effort that produced a new oceangoing ship much larger than the *caravelle*—the *galleon,* a massive vessel with huge cargo holds, the essential basis for the growth of Spain's trading empire.

Further complicating Portugal's attempt to dominate the seas was a newly resurgent England, which had decided that it also would get into the imperialism business. The English accomplished that goal very quickly because, improbably enough, one Englishman who didn't know the difference between a cube root and a turnip made some crucial technical insights, and an English scientist trying to understand how the human body worked wound up solving the most intractable problem of ocean navigation. Together, they revolutionized naval warfare and made England the greatest sea power in history.

· · ·

There were very few people who considered Henry VIII a man of deep learning. Noted for his volcanic temper, love of food (as his massive girth attested), avid devotion to dancing, and fondness for the fairer sex, the English king was a dominating figure who tended to suck the air out of any room he entered. Although he liked to boast of his disdain for books and learning, behind that hedonistic reputation was a man of deep thought. And what Henry VIII often thought about in the early sixteenth century was how he could make his small island kingdom a mighty world power. It would never approach that point by constant land wars in France that drained the royal treasury and cost English blood for a few miserable square miles of territory, such as the ruinous Hundred Years War. The answer, he concluded, was that England had to become a great mercantile and naval power, able to challenge the rising imperial powers of Portugal and Spain, who were making fortunes from overseas trade and the resources they exploited in captured lands. He especially had in mind the piles of gold and silver the Spaniards were beginning to extract from their footholds in the Americas, and the fortunes Portugal was making from its domination of the spice trade. He was also aware of the amazing voyage in 1497 by the Portuguese explorer Vasco da Gama, who sailed across the Pacific to the main Indian trading center at Calicut and came back to Portugal with a load of spices whose profits returned more than sixty times the trip's cost—the kind of return on investment that tends to attract the attention of merchants and kings. Given his devotion to the culinary arts, Henry understood all too well why da Gama had made so much money: Europeans would pay *anything* to obtain two items on his shipping manifest, sugar and tea. Put simply, the Europeans had become addicted to these two substances—addictions so strong, they influenced foreign policy for many years to come: the Dutch eventually would cede New Amsterdam (later New York) to the British in exchange for Surinam and its spice plantations, and the French would give up Canada in exchange for Guadeloupe and its sugarcane.

Henry concluded that it would be pointless for England to build a lot of merchant ships and leap headlong into the imperialism race if they couldn't be protected against predators. And the only way to do that, he insisted, was not to put cannons on merchantmen, a waste of precious cargo space, but to mount those cannons on an entirely new warship. The new kind of warship must be bigger than any previous such ship, rugged, and fast enough to keep up with smaller merchantmen. In Henry's view, the real guarantors of success for England's new maritime empire would be big warships bristling with cannons—a lot of cannons.

A tall order, because up to that point, even Prince Henry's institute had not been able to cross what appeared to be an impassable scientific hurdle: putting a lot of cannons on the decks of ships made them unstable, no matter the design. The heavier the cannons the more instability, a matter of basic physics—ships had a relatively narrow (and delicate) center of gravity easily upset if too much weight was put on the decks. The solution, also dictated by physics, was obvious: broaden the center of gravity by broadening the base of the ship, because the wider the ship, the wider the center of gravity (best understood by the difference in stability between a canoe and a wider, flat-bottomed rowboat). But widening the base of a ship exacted a severe penalty in speed, for the major factor that defined a ship's speed was its ability to slice through the water with as little resistance as possible.

Determined to overcome this technological hurdle once and for all, Henry VIII, borrowing a concept from his namesake in Portugal, organized an extensive scientific-technical effort to make England a great maritime power, including recruitment of top talent. He combined gun foundries, shipbuilders, and a whole series of supporting technical services into a government-run research and development operation that he personally oversaw. The operation's master made it clear that no expense was to be spared in creating English sea power, the main reason why a number of talented employees in Prince Henry's institute began to drift away and defy his orders by going to work for Henry VIII's higher pay. As they learned, their new boss had plenty of money to achieve what he wanted; during his dispute with the Vatican, he had seized millions of pounds worth of church lands in England, plunder that provided a lot of working capital to achieve his goal.

But Henry VIII's contribution went beyond money. When his scientists and technicians appeared stuck on the critical matter of putting guns on ships without capsizing them, Henry provided a critical insight. Why, he asked them, are you trying to put guns on the decks of ships? If the center of gravity is the sticking point, move the guns lower, toward the keel. Distribute them evenly on both sides of the keel, below the center of gravity to maintain balance. But how would they be fired? Henry's brilliant solution: cut firing ports in the hull and protect those ports from encroaching seawater by hinged, watertight doors that would be closed when the guns were not being fired.

Some authorities have suggested that Henry stole the idea from a French shipwright he had recruited, but whatever the source, the new ship that emerged one day in 1545 from the Thames to begin its shakedown voyage in the Atlantic instantly revolutionized naval warfare. The *Great Harry*, a modified *galleon*, bristled with four "great cannons" (60-pounders), a dozen "demi-cannons" (32-pounders), and a number of smaller cannons. The ship was a standoff weapon, meaning that it could remain some distance from an enemy ship and pound it to pieces with cannon broadsides. The cannons themselves were belowdecks, aimed through firing ports in the hull, just as Henry had conceived. But his scientists and technicians had added an innovation that made the ship's cannons even more formidable. The guns rested on a new type of gun carriage with small wheels that allowed the guns to roll back in recoil to a stop point that had been worked out by mathematicians after a shot was fired—which meant, conveniently enough, that the guns were perfectly sited for instant reloading by the gun crew.

Judged as a whole, *Great Harry* represented every advance the century's science could provide—hydrostatics, metallurgy, mathematics, physics, aerodynamics—and in a process that had been repeated often, in turn stimulated those scientific disciplines. This magnificent wind-driven machine was the product of a lavishly funded development effort that relied on applied science. Just to cite one example, there is no way to judge how far the science of hydrostatics would have advanced without such military development programs, but it is difficult to imagine that scientific discipline moving very far without the stimulus of a research and development program that allowed the scientists the essential components for rapid scientific development: a lot of money and a lot of time. To be sure, it was science focused strictly on a particular, applied purpose—in this case, a military one—and thus technically not "pure," but the fact was that as the various state-endowed scientific research efforts demonstrated, there was very little pure science being done. No scientist was about to get a doubloon, a pence, or an escudo from any government unless he was prepared to turn his particular scientific discipline to the service of the state—and that almost invariably involved war or the military in some form.

• • •

Henry VIII had his supership, but as every English sailor knew, it contained an Achilles heel: whoever tried to navigate it would inevitably encounter the problem of a compass that didn't seem to work too well. A lot of navigators were having the same problem, and ship captains as a result were very unhappy—as an English compassmaker named Robert Norman found out.

By 1581, Norman was practically beside himself with frustration. He considered himself the best instrumentmaker in all England, yet there was a steady parade of angry sea captains at his door, complaining that his compasses were pieces of junk: without warning, the needles suddenly deviated in certain latitudes from what the captains knew to be absolute north, such as when their ships were pointed directly at Polaris. Norman examined and reexamined his instruments and couldn't find anything wrong. He did, however, note something interesting: the reported deviations were consistent; the needles seemed to be off by consistent amounts in pretty much the same latitudes.

Norman had no insight beyond that, and in desperation mentioned the problem to a friend, William Gilbert, a medical doctor who served as personal physician to Queen Elizabeth. Gilbert was a fervent English patriot who shared his monarch's view that England's very existence depended on a great trading empire protected by a mighty navy. As Gilbert well knew, there were riches beyond belief out there, all for the taking. The vastness of this wealth had been demonstrated by Elizabeth's favorite captain, Francis

Drake, in 1580, when he returned from his famous around-the-world voyage with enough riches stolen from the Spaniards to provide merchant investors in the voyage with a 4,700 percent return on investment. Nevertheless, Gilbert realized that unless the compass problem could be solved, an English fleet, regardless of how mighty, would be just so much floating wood if it could not find its way effectively around the open seas. Drake agreed, noting that he had all kinds of problems with his compass during his voyage; he urged Elizabeth to put the weight of the royal treasury behind a major scientific effort to solve the compass crisis.

As things turned out, Elizabeth didn't have to spend a penny. She turned to the smartest man she knew, her personal physician, William Gilbert: Was there anything he could do? Gilbert was hesitant; by his own admission, he did not have deep knowledge of mathematics or physics. But he did have a first-rate scientific mind that he had devoted to trying to understand how the human body worked. He put that mind to work on the compass problem, a knotty one that took nearly eighteen years to solve. And when he did, he would create the basis for the scientific understanding of electromagnetism, a fundamental building block of modern science.

Gilbert made the conceptual breakthrough by analyzing lodestone, whose magnetic properties had fascinated scientists for many years, although nobody could quite figure out why the stone exerted a strange force upon any metallic object near it. Gilbert noted that when a compass was moved near a lodestone, the compass needle began to move. From there, he finally realized why his friend's compasses didn't work properly. In an amazing feat of deductive reasoning, Gilbert concluded that the earth is in fact a great magnet, with north and south poles. Further, the planet spins on an axis, but that spin is not entirely consistent, which accounted for the compass needle problem. As Gilbert realized, the planet's slightly irregular spin produced two norths, magnetic and "true." The deviation between the two varied, but could be predicted and measured. From there, it was a relatively simple matter to prepare navigation tables that laid out the deviations for various latitudes, allowing navigators to take those deviations into account.

While Gilbert was trying to solve the magnetic deviation problem, English sea power exploded. English privateers, covertly underwritten by Elizabeth, sailed thousands of miles to pick off Spanish *galleons* laden with treasure and sack Spanish settlements. These depredations finally led to all-out war between the two kingdoms, a struggle from which England emerged triumphant in 1588 with the defeat of the Spanish Armada. The victory was largely due to superior English technology. English scientists, in the first example of the science of hydrodynamics applied to naval warfare, built large tanks of water to test the flow of water around ship scale models to determine the best hull design. They emerged with a sleeker, faster, and more maneuverable version of the *galleon* that ran rings around the larger and less maneuverable Spanish ships.

The triumph of the small island kingdom over the much larger and richer Spanish Empire set off a new naval race throughout Europe, a search for bigger and better warships, more powerful cannons, new arrangements of sails for more speed—and, above all, improved navigation. Indeed, it became something of an obsession that would preoccupy science, especially astronomy, for the next two hundred years. But in the end, science would not solve the navigation problem. That feat would be accomplished by an English clockmaker.

• • •

In terms of getting around on the trackless oceans of the world, mankind by the end of the sixteenth century had come a long way from ancient times, when Phoenicians inscribed the famous warning on the west-facing Pillars of Hercules opposite the rocks of Gibraltar: *ne plus ultra* (nothing farther), meaning that the end of the world was situated at the opening to the Atlantic Ocean. Columbus shattered that illusion, but as he was the first to admit, navigating so vast an expanse of ocean was very much a matter of chance without the right tools. The simple fact was that navigators relied mostly on shrewd guesswork and dead reckoning to determine where they were going, an explanation for the large numbers of ships in the sixteenth and seventeenth centuries that headed out into the world's oceans and simply disappeared. But merchant ships laden with goods, and expensive warships, were too valuable to sail without an exact idea of where they were heading. Something had to be done.

Only a year after Columbus's epic voyage, Pope Alexander VI issued the Bull of Demarcation, which sought to dampen the growing tension between Spain and Portugal over the matter of overseas expansion by drawing an arbitrary line in the western Atlantic that he said was 100 leagues west of the Azores. Spain would get all land west of that mark, he decreed, and Portugal all lands east of it. That solved a geopolitical problem but created a scientific one: How, exactly, was an invisible line in the middle of the ocean to be determined? That got astronomy into the navigation business, for the only way to determine precisely where an invisible line in the sea might be was to use some sort of heavenly coordinates—which would also be handy for navigation.

Accordingly, government-funded astronomical observatories sprang up all over Europe, with the Continent's best astronomers retained at fancy salaries to search the heavens and find the predictable, immutable stellar events on which navigators could rely to steer. Their ranks included Tycho Brahe and Johannes Kepler, who would make fundamental discoveries about planetary orbits and how the solar system worked, but astronomy's real contribution to the science of navigation involved how astronomers regarded the earth. From the perspective of astronomers, the earth was, fundamentally, merely another planet whose orbit and rotation obeyed the same laws as the other planets. They discovered that the planet spins at the equator 60 miles every four minutes, which means that noon occurs four minutes later for every 60 miles of travel west, or about 1 degree of arc. The implication for navigation was enormous: a ship leaving home port could calculate its distance by knowing the exact time at home; a shipboard clock set at homeport time would reveal just how far the ship had traveled. Combined with a system that divided the globe into lines of latitude (a vertical line determined by measuring the angle of Polaris above the horizon) and longitude (a horizontal line determined by measuring the passage of the sun), both marked off in degrees, the shipboard clock promised, at last, simple and accurate navigation. But every kind of clock used aboard ships failed to work properly; none could function efficiently in conditions of a ship's motion, rolling seas, and drastic changes in temperature.

Furious work began in the European seafaring nations to come up with a reliable ship's clock. Meanwhile, renewed emphasis was put on the idea of astronomy's revelations of predictable heavenly events as the one sure navigational guide. Something of a scientific race broke out, with each country determined to achieve the greatest and most detailed astronomical observations that could be incorporated into navigational charts (e.g., precise data on the moon's phases and its position in the sky, correlated with positions on earth, would allow navigators to determine precise location).

The navigation race, especially acute between the two leading seafaring nations of the early seventeenth century, Britain and France, led to rival royal observatories—the British Royal Observatory in Greenwich, and the Paris Observatory in France. Each was lavishly funded; the Paris Observatory lured the brilliant Italian astronomer Gian Cassini to its dome by paying him the unheard-of annual salary of 9,000 livres (about $13,000 in today's money), a sum that stunned the small fraternity of astronomers, accustomed to miserable pittances they had to supplement with astrology work to make ends meet.

But to the confusion of the world's mariners, France and Britain each set its own zero meridian (the fixed origin of all latitudes). France took an early lead when Cassini's detailed observations of the transits of the moons of Jupiter (expanding on Galileo's original idea) offered navigators a reliable system of fixing positions. Its shortcomings were quickly discovered: to make the system work, navigators needed to take constant measurements of Jupiter, a task made nearly impossible by the difficulties of trying to focus a telescope from the deck of a moving ship. Meanwhile, Britain's Royal Astronomer, John Flamsteed, was concentrating on the positions of stars, part of an effort to prepare the ultimate navigation guide, a massive catalog that would tell navigators positions of stars in the sky for every single point on earth; a navigator had only to consult the positions to know exactly where he was.

Flamsteed's star charts became the preferred navigational method, and the Greenwich meridian became the world standard, as it is to this day. But the fundamental problem remained: all these astronomical guides were fine as long as the weather was clear, but any storm or severe overcast rendered all such guides useless. The only real solution was an efficient ship's clock, something that remained out of reach despite huge cash rewards offered for its successful development. It was not until the beginning of the eighteenth century, when a British clockmaker, John Harrison, came up with the insight that had eluded science for nearly two hundred years. The problem with shipboard clocks, Harrison realized, was that they all utilized a pendulum system. They were perfectly suitable for clocks resting on walls and floors of homes, but there was no way they could work on a ship in motion. To solve the problem, Harrison devised a spring-driven clock with brass slides that compensated for motion and temperature changes—the basis for all modern instruments that are subject to the same two forces. With Harrison's clock, the first true chronometer, navigation entered a golden age, and England forged ahead as the world's greatest sea power. It was a position it would not relinquish until the middle of the twentieth century.

But the real significance of the long competition to perfect navigation lies in its ripple effects. It would lead to major advances in cartography, meteorology (including the fundamental discovery that weather systems move west to east because of the earth's rotation), and surveying. Combined with advances in weaponry, these were the essential foundations for a new and terrible wave of wars that would sweep across the Continent.

It began when the people of the Netherlands decided to change the rules.

5

The Final Argument of Kings

Science without conscience is the
death of the soul.

—FRANÇOIS RABELAIS

Sometime around noon that warm spring day in 1582, Alexander Farnese, the Duke of Parma, ordered the Spanish cannons to cease firing against the Dutch stronghold of Oudenaarde, near Brussels. It would be only a momentary pause, the duke informed his gunners, sufficient for him and members of his staff to indulge in one of those grace notes that marked this Spanish nobleman as a man of culture, manners, and breeding—an alfresco luncheon in no-man's-land, just a few hundred yards from the Dutch lines.

The duke had no fear that the Dutch would take advantage of the lunch and attempt to kill the commander of the forces engaged in suppressing the Dutch revolt against Spanish rule. It was, after all, an age of civilized manners, a time of elaborate politeness and rigid social rules. This gloss of civilization had extended to war: flags of truce, mutual exchange of prisoners, elaborately polite surrender ceremonies to preserve the honor of the defeated, rules of engagement to prevent undue bloodshed on the battlefield, and attempts to shield innocent civilians from what later became known as "collateral damage" from battles. Indeed, war in Europe had become so formalized and polite, there were instances of the commanders of two opposing armies insisting that the other fire first, and truces declared for the purpose of both sides eating Christmas dinner.

So when the duke sent one of his officers under a flag of truce to the Dutch lines to convey his deepest respects to his opposite number and beg to inform him that there would be a respite in the

Spanish siege because he would be hosting a luncheon on the battlefield, he had every expectation that the Dutch commander would politely agree. He should have realized there was something wrong when his officer returned to report that the Dutch commander merely glared at him without replying, but nevertheless ordered the luncheon to go forward. A large table and chairs were carried to the battlefield, and amid the litter of shrapnel and unexploded ammunition, it was set for lunch as befitted an aristocrat—a white linen tablecloth, the best silver and porcelain, and bottles of the finest wine from the duke's own cellar. The duke and his staff took their seats as liveried servants began to bring the dishes of a four-course meal, to be highlighted by an exquisite roast pheasant the duke's favorite French chef had prepared.

The duke and his staff had just begun the first course, a French consommé, when a cannon boomed from the Dutch lines. The shot shattered the head of a staff officer near the duke, sending pieces of his skull flying in all directions; one piece struck out the eye of another officer. A second cannon shot killed two more officers, scattering their brains and blood all over the table.

The duke calmly ordered the bodies removed, a new tablecloth brought, and the table reset. As it was put in place, still more cannon shots rang out from the Dutch lines, killing another three of his officers. Finally, the duke gave up. Angrily throwing down his napkin, he stalked back to the Spanish lines, bitterly complaining to his surviving staff officers about the perfidy of these "barbarian" Dutch, who didn't seem to understand how wars should be fought.

Actually, the Dutch knew only too well how to fight a war—but not the war the duke of Parma was fighting. For the duke, the war in the Netherlands was simply another in the many dynastic and political struggles that had marked European history almost unceasingly for the past four hundred years. It was a chess game played with human pieces, and few took it personally; in a world of shifting alliances, the enemy on the other side of the battlefield today might just as easily be an ally tomorrow. French cannons were inscribed with the motto *ultimo ratio regis* (the final argument of kings), an accurate summation of how the rulers of Europe regarded war in the sixteenth century.

But to the Dutch, the war was a struggle for their very existence, a war of national liberation that also involved religion (Protestant

Dutch versus Catholic Spain). They had revolted against Spanish rule for their political and religious freedom, and they were not interested in conducting the age's typical war—brief, relatively few casualties, ponderous campaigns to maneuver an opponent out of his fortresses, perhaps a nasty battle or two, and finally a treaty that awarded one side or the other a few slivers of territory (a treaty that amounted to a brief respite until the next war). Now, however, the Dutch were fighting the first total war in Europe, a struggle to the death in which they would accept nothing less than total victory, defined as driving the Spaniards out of the Netherlands forever.

Their struggle would signal a significant change: European war was about to be transformed into great, terrible struggles to achieve total victory. Over the next two centuries there would be fewer than ten years of peace in Europe as wars become more frequent, more violent, and more bloody. "This is the century of the soldier," the Italian poet Fucio Testa said in 1641, and he was right: armies became larger, their weapons became more powerful, their costs skyrocketed, and the damage they wreaked became more extensive. From that bloody day at Oudenaarde to the end of the eighteenth century, there were dozens of bloody wars, including the Thirty Years War, which totally devastated Germany from one end to the other; an attempt by one ruler to take over the entire Continent; two political revolutions; and the first world war in history.

There were a number of reasons for this transformation—the rise of nation-states, deepening religious antagonism, nationalism—but the chief reason why war became total was that the men who conducted war now had the tools to do so totally, tools science provided. It is no coincidence that the golden age of war occurred at the same time as the golden age of science, for it was science, harnessed to politics, that formed war's cutting edge. War's insatiable demand for bigger and better ways of killing men and the search for the silver bullet that would guarantee total victory were the chief impetuses for the scientific revolution that transformed the world. The main building blocks of that revolution—precise time measurement, enhanced astronomical observation, great technical advances in navigation, insight into the properties of chemical substances, the mathematics to explain and predict natural occurrences—were almost entirely spawned by military considerations.

How that intertwined process came to be is best illustrated by the bloody course of events in the Netherlands and the sobering story of an ambitious young king who liked to dance in a sun costume.

• • •

Like almost all of his fellow Dutchmen, Prince Maurice of Nassau was a fierce patriot obsessed with the idea of a Netherlands free of Spanish domination. His formal title was Captain-General of Holland and Zeeland, which meant he was head of Dutch forces in the United Provinces, the territory of what is today the northern part of the Netherlands and western Belgium, trying desperately to hold out against strong Spanish pressure to bring all of the Netherlands under the Spanish crown. Maurice faced a formidable task, for Spain, flush with wealth from its New World plunder, had built a formidable military machine, the best in Europe. Its greatest strength were the *tercios*, the highly trained infantry armed with the *arquebus* and the early muskets; they functioned as a well-oiled machine, with ranks alternately kneeling to fire while other ranks carried out the complex loading procedures for the weapons, then switching the ranks from loaders to firers after the first volley.

Maurice, a university-trained mathematician, was aware that the lesser-equipped and poorly trained Dutch forces were no real match for the Spaniards, so he began thinking of an edge that would redress the balance. He began trolling among Dutch-controlled territories for scientists who could come up with innovative ideas and wonder weapons, finally emerging with a real gem: Simon Stevin.

Something of a late bloomer, Stevin was a former bookkeeper and tax clerk who had entered the University of Leiden in 1583 at age thirty-five to study science—and immediately found his true calling. Stevin first demonstrated his brilliance in mathematics, revolutionizing computation by devising the use of decimals to replace the practice of calling all partial numerical terms fractions, allowing them to be included in positional notation. He also devised rules for locating the roots of equations. These innovations caused a sensation in the world of mathematics—and the enduring gratitude of mathematicians who routinely faced such computational tasks as $3/122 \times 4/651 - 1/73 = ?$ With Stevin's decimal system, such tasks are routinely handled today by elementary school students. (Stevin's work with decimals led him to the bright idea of using them for coinage, a system that would make accounting and commercial transactions much simpler to perform. He found no takers for the idea, but many years later, an American diplomat in Paris, Thomas Jefferson, heard about it and successfully proposed its use for the currency of the new United States.)

Stevin's feat brought him to the attention of Maurice, who enlisted him in the war effort as his scientific adviser. Like his patron, Stevin was a fervent Dutch patriot, and so saw no problem with turning his considerable scientific talents to war. He proved his worth almost immediately by inventing an amphibious boat in which Maurice could dash around to inspect Dutch seacoast fortifications. And those fortifications soon underwent a dramatic improvement when Stevin came up with an entirely new system, a complex geometric layout that created deadly interlocking fields of fire, along with a series of moats and ditches that kept besiegers at bay. The system was constructed with the help of new advances in trigonometry Stevin devised.

But Stevin's real contribution to the Dutch war effort was a secret weapon: water. He had spent a great deal of time researching the science of the factors determining the pressure of liquids on surfaces. Today known as hydrostatics, the branch of physics that deals with liquids at rest, it was virtually invented by Stevin, who was inspired by his studies of Archimedes (he of the famous bathtub incident). Searching for something that would checkmate the Spanish armies, he noted that a good part of the Netherlands had been reclaimed from the sea, which was held back by a series of barriers. Stevin's study of how the force of water acted against those barriers led him into a crucial insight, one that underlies all hydraulic systems, hydroelectric dams, and vehicle braking systems today: the pressure exerted by a liquid upon a given surface depends on the height of the liquid and the area of the surface. Moreover, that pressure could be mathematically deduced, meaning that it could be controlled. Armed with that insight, Stevin had a series of sluices built, through which huge amounts of water were unleashed when necessary to instantly flood a lowland plain on which a Spanish military force happened to be standing.

When the Spaniards devised large pontoon bridges as a countermeasure to Stevin's flood weapon, he quickly came up with a counter-counterweapon, the world's first guided missile. In 1590 the Spaniards had constructed a huge pontoon bridge over flooded land to besiege Antwerp. Working with a group of Italian naval engineers Maurice had recruited, Stevin supervised the construction of a floating time bomb—an ordinary merchant ship whose hull was lined with bricks and packed with gunpowder and scrap iron. After arming it with a fuse he had mathematically calculated to ignite the explosive mixture at just the right time, the ship was pushed toward the pontoon bridge, which was packed with thousands of Spanish troops. They paid little attention to the small ship slowly drifting with the current toward them, just another ship in a port city busy with such traffic. As it reached the bridge, the ship exploded in a huge blast that scattered wreckage a mile in every direction. More than two thousand Spanish soldiers were killed outright, the greatest number of casualties caused by the discharge of a single weapon in history to that point. The disaster caused shock waves in Spain, whose determination to subdue the Netherlands now underwent a crisis of confidence: How was it possible to defeat people who flooded their own land and used terror weapons? Finally, six years later, Spain withdrew from the Netherlands, a decision hastened when England and France recognized the newly proclaimed Dutch Republic.

Although big-power politics played a role in the final resolution of what Spain had come to call "the Spanish ulcer," the fact is that David had brought Goliath to his knees. And the central lesson could not have been more clear: The Dutch ultimately triumphed because they had superior science, most notably the genius of Simon Stevin. The time of total war had dawned; future wars, like the struggle in the Netherlands, would involve all the people and resources of a warring state. In a time of rapid scientific and technical change, the greatest resource would be science, the resource that had humbled the great military power of Spain. For that reason, Europe paid careful attention to the long list of scientific and technical innovations contributed by Maurice and Stevin to the art of war, new tools that remain in use to this day—operational research, utilizing lead soldier figurines to deduce improved tactical ideas (giving birth to a hobby that has lasted centuries); schools to train a new technical class called military engineers; the importance of scientific training for senior officers; the vital importance of scientific advisers to rulers; military manuals to disseminate new technical know-how to every level of military forces; and an ongoing, large-scale research and development program to produce a constant supply of new and improved weapons systems.

These ideas would take their firmest root in France, where the man who would dominate—and terrify—much of a century wanted his nation to become the supreme power in the world, a status he meant to achieve by force of arms.

• • •

For a man who aspired to be the next Alexander the Great, Louis XIV, king of France, did not cut an imposing figure. Diminutive, he wore red high heels to create the illusion of greater height, and often had himself painted in various heroic or martial poses, although in fact he was almost totally inexperienced as a soldier. His real passion was dancing, which began early in his life. At age fifteen, he wrote an original ballet, *Ballet de la Nuit*, a thirteen-hour-long bore that nevertheless excited rapturous enthusiasm among audiences at the royal court, although their lavish praise for the work undoubtedly had much to do with the wish to get on the right side of the heir to the French throne. The enthusiasm extended to the lead dancer, Louis himself, who performed the role in a sun-motif costume, a garish outfit that led to his later nickname, "the sun king."

But that nickname had a double meaning, for it also referred to a passion of Louis even more consuming than dancing: power. From the moment he ascended to the throne in 1661, Louis made it clear he intended nothing less than to make France the master of Europe. France, he said, would be expanded to its "natural boundaries," which he defined as the Alps, the Rhine River, the Atlantic Ocean, and the North Sea. To achieve that goal, he decreed, France would fight whomever it had to fight (including non-French people who happened to occupy land Louis coveted) and spend whatever had to be spent. Harsh, egotistical, and an absolutist who believed that like all kings, his right to rule was divine, handed down directly from God, Louis made this momentous decision himself—and further made it clear he alone would dictate whom France would fight, and when. It was an approach summed up in his famous reply when someone had the temerity to question any of his decisions: *"L'état, c'est moi!"* (I am the state!).

As a result, France was constantly embroiled in war for almost all of the next fifty-four years of his reign, a record of bloodshed and strife barely equaled since the days of the Roman Empire. "I have loved war too much," Louis would lament on his deathbed in 1715, by which time France had risen to become the most powerful and prestigious country in Europe, the world leader in both the cultural and military arts, a glittering jewel whose heritage is part of all Western culture. His success, however, had a price: France was virtually ruined in the process, creating the conditions for the revolution that later would destroy the French monarchy forever.

Among the more glittering achievements of Louis' reign was a great flowering of science, and that was no accident, for it was science that played a vital role in France's rise to greatness. Louis, barely literate scientifically, nevertheless was a devoted student of power politics, most especially the dramatic course of events in the Netherlands some years before. Louis concluded that the Spanish defeat at the hands of "a flock of tulip-growers," as he contemptuously termed them, was due to the Dutch superiority in science and their ability to devise new wonder weapons and techniques to balk a militarily superior enemy. To dominate Europe, he concluded, France must be the most scientifically advanced nation on the Continent, and science must be organized and directed to serve the needs of the state exclusively.

Fortunately for Louis, he had just the right man at hand for the task: Jean-Baptiste Colbert, officially finance minister, although his mandate extended far beyond that job title. An exceptionally brilliant administrator, Colbert was put in charge of a well-funded effort to create French scientific dominance overnight, a task Louis hoped to accomplish with lavish offers of gold for the cream of scientific talent he sought to recruit. To a large extent he succeeded, although some of the scientists he enrolled later would have cause to wonder just what they were doing in France.

Colbert believed that the first step toward French greatness would be to build a naval power capable of challenging England's dominance of the seas. As Colbert told Louis, the chief reason for the English dominance was its Admiralty, which had organized an extensive research and development operation that kept England in the forefront of naval science. For that reason the English had the most advanced ships, the best guns, superior navigation, the best shipwrights, and the best naval officers (the products of naval train-ing schools, which emphasized technical and scientific training for future naval officers).

Colbert set about transforming France's minuscule navy into a first-class power, moving on two fronts simultaneously. First, aware that efficient navigation was the primary key to naval power in the seventeenth century, he sought to make France the center of world astronomy, the scientific discipline he believed held the answer to navigational problems. He convinced Louis to invest several mil-lion livres of the royal treasury in the state-of-the-art Paris Conser-vatory, to which he lured the cream of European astronomers, most prominently the Italian astronomer Gian Domenico Cassini. The astronomers were under orders to devise the most detailed possible observations of planetary and star movements with the kind of pre-cision that would allow navigators to use them for pinpoint star fixes. They were also told to determine the precise shape of the earth so that navigators could more precisely fix their positions.

Meanwhile, Colbert enlisted the other sciences for his naval buildup. He established schools of hydrography and hydrology, the sciences of charting water-covered areas of the earth, and the physi-cal and chemical composition of the planet's waters, respectively. He also recruited a small army of foreign naval engineers, scientists, and shipwrights under a state-controlled research and development effort to come up with new and improved ship designs. Emulating the English example, he established naval academies and technical training schools for future naval officers, along with a state school of naval architecture.

Within ten years Colbert's efforts converted a tiny, virtually im-potent navy of 18 outdated ships into a powerful, modern force of 190 ships staffed by highly trained naval officers. At the same time, he worked to make the French army the best in Europe, using many of the same innovations that had rebuilt French naval power—military academies to produce scientifically and technically literate officers, extensive research and development centers to keep France on the cutting edge of the latest science, arsenals staffed with scien-tists to develop better cannons and small arms, and military re-search institutes to study the latest developments in strategy and tactics. The result was a standing army of four hundred thousand men, an enormous force by the standards of the day, equipped with the most up-to-date arms (including a French invention called the bayonet, named after its first use at a battle near the town of Bayonne). To make this vast force mobile, Colbert turned much of France into a military logistical system, rebuilding country dirt tracks into a network of first-class roads over which troops and their supply trains could be moved quickly, dredging rivers to make them navigable for warships, and building France's famed network of canals (still used today) to provide rapid waterborne transport for military supplies.

While this great bootstrap effort was transforming France into a mighty world power, it also caused a great flowering of a host of scientific disciplines, among them cartography, a discipline with immediately obvious military implications. Up to the time of Louis, cartography was very much an inexact science; most maps were notoriously inaccurate, since they did not rely on scientific surveying techniques. Louis decided that if he was going to conquer Europe, he better know precisely what his borders were—and everybody else's. To that end he recruited a topflight group of mathematicians, astronomers, and geographers, and commanded them to use science to achieve the unprecedented: absolutely accurate maps on which he and his military could rely. In other words, if a map said that point A was 7.5 kilometers from point B, the map's reliability on that point had to be unquestioned; similarly, if a map said that a particular hill was precisely 145.6 meters high, that altitude had to be absolutely correct. The goal, Louis said, was to allow France to locate every single place on the face of the earth scientifically.

Using the talents of astronomers who fixed precise star positions to locate precise points on the ground, geographers who accurately mapped land features, and mathematicians who used the new scientific instrument of the telescope to accurately measure distances, the group was able to produce the first truly accurate maps in Europe—which proved immensely valuable when Louis' armies went on the march. The first examples of these maps were presented with some trepidation to Louis, since it was discovered that the newly accurate measurements had reduced the king's presumed land holdings by some 20 percent. "Excellent, gentlemen," Louis remarked dryly as he was shown the reduced borders of his kingdom. "I had no idea that when I enjoined you to accurately survey, I was asking you to take land from me."

Cartography as a modern science was born in this effort, as was forestry. That came about because while Colbert was frantically rebuilding the French navy, he was denuding the French forests of oak, the hardwood used to build ships (about a thousand oak trees were required to build one ship). An alarmed Louis hectored his scientists: either find a substitute for oak, or figure out a way for France to have an inexhaustible supply of oak. The answer came from several horticulturists, who, after careful study of the life cycle of trees, concluded that forests were a renewable resource; with proper planning that regulated exactly how many trees were to be felled each year, enough new trees could be planted to guarantee a steady supply.

In terms of impact, no scientific discipline was more affected during Louis' reign than chemistry, a circumstance that grew out of military necessity. There was, for example, the great soap crisis, a by-product of Colbert's massive military buildup. That buildup was using a lot of wood, which left hardly any for such things as making soap, at the time produced from alkali, which in turn was produced by wood ash. Louis ordered the soap industry to stop using wood ash to help stop the wholesale consumption of France's trees, then ordered his scientists to come up with a substitute; an unclean army in dirty uniforms was unthinkable. Experimenting with various substances, the scientists finally hit upon the idea of burning kelp—a seaweed in plentiful supply along France's Atlantic coast—whose ash made perfectly fine soap. (In the process, they discovered that processing kelp produced a wondrous chemical by-product called iodine, which French military surgeons found useful for treating infections and cleaning wounds.)

More importantly, French chemists were assigned the task of improving gunpowder, especially a process that would produce more of it at cheaper cost (Louis had envisioned using a lot of gunpowder). This essential military substance was still relatively expensive to produce, and worse, contained chemicals, especially sulfur, that France had to import—and that might be cut off in the event of war. After several years of work, the chemists hit upon cheap components for making gunpowder, which involved recovering saltpeter from the efflorescence on walls and combining that with nitrogenous animal wastes. They also solved the sulfur problem by figuring out how to retrieve sulfur from volcanic and other deposits, relatively plentiful in France. For good measure, they devised new grades of gunpowder that could be used for different-size weapons, including a "super charge" used to propel cannonballs from the biggest guns.

But by far the biggest impact was something that no one had anticipated—or even named yet: the industrial revolution, the single development most responsible for making war truly total. The revolution began during Louis' reign when, armed with the new, cheaper gunpowder process his chemists had given him, and eager to produce as much of the stuff as possible, he decided on a single, centralized manufacturing system. The idea was to avoid the production delays inherent in the traditional system, which involved a far-flung network of individual arsenals that produced varying amounts of gunpowder by hand. Louis ordered the building of a vast workshop in Paris to produce gunpowder. Quite unwittingly, he created the industrial revolution, because in the process of trying to come up with a way to make as much gunpowder as possible in the shortest possible time, the chemists discovered they would have to set up an entirely new production system. That involved, under one roof, laboratories to produce the raw materials, an array of tubs to mix the ingredients, copper boilers to heat the mixture, drying rooms for the final product, and assembly areas to pack it into wooden kegs. The system devised to coordinate all these activities on a virtual assembly line became the basis for the later revolution in manufacturing processes.

If he returned to earth today, Louis undoubtedly would be surprised to learn that he had created the industrial revolution, and absolutely dumbfounded to learn that he had taken the first step into the computer age. That came about because he wanted nice blue and white uniforms for his mighty army, a requirement that put a tremendous strain on the French textile industry. As with other industries of the time, it was essentially a handicraft, with individual looms turning out cloth. To fulfill Louis' requirements, the textile manufacturers had to come up with some kind of new system that would shorten the time necessary to make cloth and turn it out in greater quantities. The solution was an ingenious system of perforated wooden boards coded to settings on the looms; to turn out a particular pattern, it was coded in the form of perforations on a board, which then directed the loom to produce that pattern precisely. It was the forerunner of the punched card system of early computers. (He probably would be less astonished to learn that virtually all modern military drill comes from his army, which was trained to operate with clockwork precision on command and to fire by the numbers. Essentially, the army was choreographed, the product of efforts by Louis' dancemaster, Pierre Beauchamp, known better for his creation of classical ballet, which uses the basic movements he invented and is known still by the names he gave to them, such as *pas de deux.*)

The cost of all this scientific innovation was enormous, and Colbert constantly fretted about where France would get all the money. Louis seemed to have no grasp of economic reality, and to Colbert's dismay, insisted that France could have guns and butter at the same time. Of course, he meant butter for himself and the royal court, not the ordinary French citizens bowing under the weight of high taxes. And that court, in terms of sheer luxury and extravagance, was like none other.

Nobody seemed to know just how much Louis' palace at Versailles cost, but as Colbert calculated, it must have been astronomical. Even worse, at a time when he was trying to scrape up every franc he could find to build a great navy and fund the huge standing army of four hundred thousand men (not to mention the heavy costs of all the wars Louis was precipitating in the process of grabbing territory), the scale of excess at Versailles was nearly beyond belief. It was a fairy-tale world of mirrors, gilded walls, and marble set among hundreds of acres of luxuriant gardens, pools, and fountains that required a small army to maintain. Louis meant this extravagance, the most luxurious in the world, as a glittering reminder of the splendor of a newly resurgent France, the kind of glitter where visiting diplomats would encounter a French king in a magnificent robe made especially for diplomatic receptions—an indescribable garment of lace and gold bordered with diamonds that an appalled Colbert learned had cost the treasury some 12 million livres (about $20 million). The upkeep on Versailles also ran to astronomical sums, and no wonder: at any given time, Louis might have a thousand aristocrats staying at the huge palace, where they passed the time in grand balls, formal dinners, gambling, or occasionally gamboling in that new French invention, the swimming pool. These aristocrats, who wouldn't even think of traveling to Versailles without a minimum retinue of eleven servants, lived in their own universe without any conception of such matters as cost; one of Louis' mistresses lost some 2.5 million livres (about $4 million at current rates) in one night of gambling, a loss she simply laughed off, secure in the realization that there was plenty more where that came from.

Whispers about the excesses at Versailles began to spread to France's scientific community, causing many to wonder: Louis had enjoined them to come up with the kinds of innovations that, among other things, would allow France to produce things cheaper and better, yet there he was at Versailles, bleeding the state treasury as though it were bottomless. If the purpose of their work was to make life at Versailles more fun for the aristocrats, that was not why they were sweating over laboratory benches. One of them, the distinguished French astronomer Jean Picard, was summoned to Versailles to discuss "an important scientific matter" with the king himself. Believing he was about to discuss some weighty topic, perhaps involving the shape of the earth's poles (French astronomers were working hard to determine whether they were oblate-shaped or round, a question with implications for navigation), Picard showed up at Versailles laden with documents and diagrams.

But to his shock, Louis quickly made it clear to Picard why he had been summoned, and it had nothing to do with science. He pointed to the palace gardens and told Picard that the various water amusements—fountains, pools, and statuary that spouted water (statues of little boys urinating seemed to be a dominant motif)—didn't seem to be working properly. Was there something Picard could do?

An irked but nevertheless loyal Picard immediately dropped such important tasks as his researches into the eclipses of Jupiter's moons and dutifully went to work in Versailles to solve the great water crisis. In short order he discovered it was a simple problem: Versailles was on slightly higher terrain than the surrounding areas, where the cisterns and reservoirs that fed the water sports were located. Picard had the cisterns and reservoirs elevated, and soon the water was flowing at full power once more.

Louis was delighted and escorted guests around the gardens, showing off the fountains and statuary that gushed freely again. There was, however, one curious incident: person or persons unknown, for some odd reason, had thrown a small brass cannon into one of the swimming pools. Whether it was done by Picard in a fit of pique has never been determined, nor is it known what message the miscreant may have been attempting to send. A protest against the excesses of Versailles? A subtle jibe at Louis' insistence on having vast military expenditures and luxury at the same time? A humble French workman's protest at the burdensome taxes he was paying to support cannons and swimming pools, neither one of which did much to feed his family?

No one knows, although several of the aristocrats found it all highly amusing; some sort of interesting new diversion, they supposed.

• • •

Louis XIV's great leap forward scientifically and militarily caused tremors of anxiety in France's powerful neighbors, especially her most powerful Continental rival, England. The English were only too aware of the importance of science in maintaining their powerful position in international politics, and were determined to keep it. That determination resulted, in 1662, in the formation of the Royal Society for Improving Natural Knowledge, a combined scientific society and think tank underwritten by royal funds. The official purpose of the society was to act as a sort of clearinghouse of scientific research in England. Society meetings featured member scientists (called "fellows") discussing their latest findings or research, which were published by the society in the form of reports, the prototype for the scientific research paper.

Although the ninety-eight charter members of the society liked to consider themselves "pure" scientists dedicated only to expanding man's knowledge of the workings of nature, they were also English patriots. Moreover, they needed no reminder that the king of England paid the bills, so the Golden Rule came into effect: the society was soon bombarded with official requests to address various technical and scientific problems that were strictly military—navigation, smaller and lighter cannons so that more guns could be packed into warships, safer gunpowder, more powerful guns. They also needed no reminding that France, in addition to its new navy, also had been drastically overhauling its army, a vast force equipped with improved weapons designs and relentless drilling under a ruthless taskmaster named Martinet (from whom the term meaning rigid taskmaster is derived). That army now featured battalions of soulless robots who marched into battle at a rate of precisely ninety paces a minute. And once on the battlefield, this superb force could form unwavering ranks to deliver volleys of fire with mathematical regularity. There was also the growing threat from the burgeoning military power of Prussia and its army, a disciplined force that made the French army look like a mob (the fabled Prussian discipline would later reach its apex under the Frederick the Great, who once stunned a group of foreign military observers at his maneuvers when on his command twenty-three regiments of Prussian infantry, some twenty-four thousand men, instantly and flawlessly wheeled into line from a marching column).

Ever since the Spanish Armada, the English had been acutely aware that their small island kingdom faced constant danger from larger rivals, and that as a seafaring nation, their vital sea lanes were always in danger of being cut by any power with superior naval technology. England had managed to keep out of imminent danger, a circumstance largely attributable to its ability to harness its best scientific brains to war. Not all the scientists were unabashedly patriotic about using their knowledge for such purposes; witness the case of one of England's greatest mathematicians, John Napier.

A devout Calvinist, Napier was a mathematician who preferred to devote his efforts to pure mathematics (his invention of the logarithm would revolutionize computation), but when the Spanish Armada neared England, he was asked to put aside his mathematical work and design "war machines" for his homeland. He helped naval designers come up with the new, fast English galleon that devastated the Spanish fleet and saved England from an invasion, but some years later, when he believed the threat to England from Spain was finally at an end, he declined to do any further "war work." Only after his death in 1617 was it discovered that like da Vinci, he had been filling notebooks with all kinds of amazing weapons designs, including the submarine and the tank. Most incredible of all, however, was what he described only as a "weapon of mass destruction" that could devastate every living thing in an area of 4 square miles. Napier was deliberately vague on just what that weapon was, but some experts have suggested it may have been some form of nuclear bomb; Napier apparently had begun to realize the tremendous destructive power inside the atom. Whatever it was, Napier refused to reveal anything further, even when, on his deathbed, he was visited by a friend and urged to tell him the secrets of the weapon that would make his beloved England militarily invulnerable forever. No, Napier replied; there were already "too many devices for the ruin and overthrow of man," then passed his last breath, taking the secret to the grave with him.

No such tugs of conscience afflicted Napier's mathematical inheritor, the equally brilliant Benjamin Robins, although he had been born of Quaker parents. A fervid English patriot, Robins left full-time mathematics to devote his energies to military engineering, believing that unless England remained at the forefront of applied science, it had no hope of surviving among a pack of hungry wolves. The key to that lead, Robins believed, was gunnery, and it was there that he applied his considerable brainpower. Recruited by the British army's Ordnance Department, which ran a flourishing research and development program to improve its guns, Robins focused his interest on interior ballistics (the motion of projectiles inside guns) and terminal ballistics (the behavior of projectiles at the end of their flight). Extensive experiments with all types of gunpowder and guns, ranging from giant cannons to muskets, led Robins to a series of groundbreaking mathematical equations that for the first time spelled out exactly the forces that propelled projectiles out of guns and the muzzle velocity as they left the barrels. These proved to be essential tools for the design of newer and more powerful guns, especially the *carronade,* a short, lightweight gun built to Robins's specifications and that could pack a lot of punch for its size. It was ideal for a new class of English super warship known as a "ship of the line," bristling with up to eighty guns.

Robins's contributions didn't stop there. He concluded that English gunners were using too much powder in their guns, an error he solved by devising what he called a "ballistic pendulum," a calibrated instrument that measured muzzle velocity so that gunners could achieve just the correct amount of propellant charge, depending on what range they wanted to achieve. But however efficient Robins made smoothbore guns, his equations revealed a definite limit: such guns would achieve only a certain upper range of efficiency and range, mainly because they were smoothbore in the first place. Given the physics of what happens to explosive gases inside a hollow tube, Robins realized, even the most efficiently produced gunpowder and ammunition manufactured to the most precise specifications would never achieve really devastating hitting power and very long range as long as they had to operate in that kind of tube. Robins's solution to the problem would revolutionize gunnery and in the process create an essential tool to make war much more deadly.

Robins's experiments, in which he painstakingly examined every moment in the flight of a projectile fired from a gun, led him to a brilliant insight: projectiles achieved their greatest range and hitting power if they were spinning while in flight because spin helps overcome air resistance. But how to make them spin as they exited the gun? The answer, Robins believed, was to make the barrels grooved in a way that took the projectile at the beginning of its flight phase— as it began to move forward from the force of the explosive gases behind it—and spun it while it was still moving up the barrel. Tests confirmed the hypothesis: a "rifled" barrel caused a projectile to emerge from the mouth of the barrel with more than five times the muzzle velocity of smoothbore guns. Moreover, the projectile, backed by greater impetus, traveled much farther and with greater accuracy.

The next step was to determine the best aerodynamic shape for a spinning projectile. That turned out to be a concept Robins borrowed from the design that made the fifteenth-century longbow arrow such a devastating weapon: a cone-shaped warhead that came to a sharp point. Finally, another conceptual breakthrough: breech-loading guns, meaning guns with a breech that could be opened, a powder charge and projectile inserted, then closed with a screw mechanism that sealed the firing chamber at one end. That would create an explosion of maximum efficiency. Such firing chambers, attached to an elongated, rifled barrel, would generate a very high acceleration rate for the spinning action of the projectile.

All this represented a remarkable scientific tour de force by Robins, striking achievements in thermodynamics, aerodynamics, and mathematical physics that created the science that made possible the terror of modern artillery and military small arms. Their power would not reach full fruition until several more decades passed, when technology could catch up to Robins's ideas—breech-loading artillery with long barrels to create maximum spin; projectiles with powder charge and warhead in one unit grooved to catch the rifling in the interior of the barrel; and new metallurgical techniques to enable artillery to stand up to high temperatures and repeated interior explosions. When that moment arrived, artillery became truly terrifying, a weapon capable of wreaking awesome destruction: high-explosive shells that could flatten an entire building from overpressure (higher atmospheric pressure created by the explosion); 100-pound shells that could dig a crater several feet deep; 16-inch rifled naval guns that could shoot a shell 20 miles; shells whose iron casings would shatter on impact into hundreds of thousands of steel slivers to slaughter any troops caught in the open within 400 yards in every direction. The standard infantry musket, converted into a true rifle firing cone-shaped bullets instead of musket balls, also became a powerful weapon: an infantryman could hit a target up to a mile away, and a copper-jacketed bullet could penetrate solid walls up to half a foot thick.

Robins's work put England in the forefront of weapons development, a lead that increased dramatically when the British were able to link their technological breakthroughs to the full flowering of the industrial revolution. That revolution had begun in France when Louis XIV wanted to increase the production of gunpowder, a technique soon emulated in the gunpowder mills of Britain, Germany, and Russia (which had 633 workers in its main powder production facility, using water-powered machinery). France would have remained the leader in industrial production except that Honoré de Blanc encountered a political problem.

Considered France's finest gunsmith, de Blanc in 1750 approached the government with an idea: standardize the production of muskets. The problem, he pointed out, was that the great European arms race was causing severe production problems in turning out muskets for the increasingly large armies, especially in France. The weapons were produced, one at a time, by individual craftsmen, but that meant if a part broke, it could be replaced only by another part produced by the same craftsman who made that individual gun. The solution, de Blanc said, was to emulate France's gunpowder industry by setting up a similar operation to turn out muskets. All the muskets produced in the process would be exactly the same, and thus their parts would be interchangeable. The French arms craftsmen, convinced that their jobs would be ended by such a process, protested vehemently to the government, threatening to stop working on muskets if a musket factory were built. Fearful of losing their services, the government backed down and French musket production went on as before, despite complaints from the French army that it was not getting a sufficient number of guns.

The muskets made by the craftsmen were perfectly fine, but the same cannot be said of French cannons, which were experiencing severe problems. The quality of the cannons tended to vary widely, depending on which particular arsenal made them, and by 1773, the French army was demanding that the government do something. Too many of the cannons were blowing up in the crews' faces, leading army cannoneers to say that the French army had more to fear from its own cannons than from enemy guns. The immediately obvious solution—set up a central manufacturing facility, as had been done with the making of gunpowder—ran into the same roadblock as de Blanc encountered: the cannon craftsmen threatened to walk if such a thing came to pass. But instead of caving in, as they had with the musketmakers, this time the French government decided that only some kind of radical solution would solve the problem, an entirely new way of producing cannons that would remove the manufacture from the hands of the craftsmen. A committee of scientists, engineers, and ordnance experts was set up with orders to devise a solution—and a quick one.

Fortuitously enough, the committee had just gotten down to business when some of its members heard whispers about a revolutionary new process in England for making cannons, a process that actually mass-produced them. Even better, the process made cannons with interchangeable parts. Considering the military implications of such a technological breakthrough, it would seem likely the process would have remained a closely guarded secret. But, as the French discovered to their delight, the man who invented the process, although he considered himself a patriotic Englishman, also happened to be very money hungry.

John Wilkinson, the best ironmaker in England, was so obsessive about the subject of metal he slept with a steel ball in his hand so that if he dreamed of a great idea and his body twitched, the ball would fall to the floor, waking him up, and allow him to jot down whatever brainstorm had come to him in his sleep. He had come up with a number of technological innovations that dramatically improved the quality of the cannons and the muskets of his main customer, the British military, but felt the government was insufficiently compensating him for his work. And compensation was no small matter to Wilkinson; his vision was to build the biggest and most profitable ironmaking enterprise in the world and thus become very wealthy. It was in this resentful frame of mind that he met a visiting French delegation that cautiously broached a delicate subject: Did Wilkinson have anything at hand that would solve the French cannon crisis? And, the Frenchmen added hastily, if Wilkinson just happened to have a suitable innovation he'd be willing to sell, France would pay very handsomely.

He certainly did: Wilkinson proudly showed off his latest innovation, a water-powered cylinder boring machine that could hollow out a cannon barrel from a solid piece of cast iron. True, Wilkinson admitted, the process was slow, but the advantages were obvious: solid casting meant a much stronger cannon, better able to resist the stress of explosions. Even better, the process guaranteed that each cannon would be exactly alike. Once the boring machine had been properly set to make one cannon, every one made by the machine thereafter was a precise duplicate. The French instantly made a deal, and arrangements were made to begin shipping Wilkinson-bored cannons across the English Channel. Something of a tricky proposition because, as Wilkinson and the French were aware, the British government was not about to let this technological edge slip away. Although France and Britain were not at war in 1774 (one of the few times in the century when they weren't), the British government was very sensitive about any of its cutting-edge military technology leaving the country for elsewhere, including France. Accordingly, British Customs officials examining ship cargoes were presented with documents attesting that the merchant ships stuffed with all those pieces of iron that bore an uncanny resemblance to cannon barrels were, as the documents insisted, "iron pipings." Fortunately for Wilkinson and the French, the Customs agents did not bother to wonder why France, perfectly capable of making its own piping, needed to come all the way to England to buy British iron pipes.

This early technology transfer solved the French cannon crisis, making Wilkinson a rich man. He was about to get even richer, because while he was wondering how he could increase the value of his innovation by increasing the power of his boring machine (and thus turn out more cannons faster), James Watt walked into his life. Watt, a polymath and inventor, had come up with the radical idea of what he called a "steam pump engine." Although Watt claimed he had gotten the idea while watching the water boil in his mother's tea kettle, in fact he was inspired after seeing a crude piston and cylinder device pump water out of coal mines. Watt had built an advanced version, but as he told Wilkinson, the problem was that the machine needed high-precision pistons and cylinders to work properly, and that was something nobody seemed capable of providing. Wilkinson dove into the task, emerging with a new form of crucible steel from which he made pistons and cylinders with tolerances "ere the length of an old shilling," as he put it. Now Watt's steam-powered machine worked efficiently, and the industrial revolution took off like a skyrocket.

In military terms, that revolution, combined with the continuing advances in the power and range of guns, now fulfilled the key scientific-technical requirement that made possible mass armies with massive amounts of arms. Arms factories powered by Watt's steam engines could turn out a huge number of guns and ammunition, all precisely the same as the prototype—which in the case of guns also meant interchangeable parts. Another requirement was communication: How to control and communicate with all those troops? The answer to that problem came in 1794, when a French military engineer named Claude Chappe invented a system to handle communications between France's main military headquarters in Paris and outlying field headquarters. As Chappe realized, the growing size of armies and their logistical requirements created a critical need for some kind of new communications system to replace the hopelessly slow method of dispatch riders and couriers, all limited by the speed of a horse. His solution was a series of towers, spaced about 10 miles apart, each one of which contained movable beams that could be set in up to ninety-two different positions to convey letters, words, or entire phrases. Each tower would relay a message from other towers sequentially, read by telescope; when the French signalmen mastered the system, a message could be dispatched at a speed of about 400 miles a day.

Alarmed by this French technological breakthrough, the British instituted a crash program to match it, resulting in a version of the Chappe system that used a wood frame with five shutters that were opened and closed in coded patterns, read by telescope, and then relayed farther. The system was used to convey orders from the

Admiralty in London clear across the country to the main naval base at Portsmouth in minutes. (Almost simultaneously, several British scientists, familiar with the early experiments on the mysterious force known as electricity, proposed the idea of what they called an "electro-optical" system for military communications. None of them could figure out how to harness the new phenomenon of electricity for the task, and it would not be until 1840 that a nonscientist, an American tinkerer named Samuel Morse, finally brought the idea to fruition.)

With mass production and more rapid communications in place, all the requirements for mass warfare were now ready. All it required was a military leader who understood how to use them. Just as the British were putting the finishing touches on their new naval communications system, that man suddenly and dramatically burst onto the world stage and became the greatest conqueror since Alexander the Great, in the process redrawing the map of Europe and revolutionizing war.

By any standard a remarkable career path for a humble, Corsican-born French artillery officer who liked to call science "the first god of war." But the relationship between science and Napoleon Bonaparte would prove to be a rocky one—to his detriment.

6

Prometheus Unchained

*Science belongs to humanity in peacetime
and to the fatherland in war.*

—FRITZ HABER

He was a diminutive man, but Nicholas Appert confidently faced the distinguished members of the Society for the Encouragement of French Industry seated around a large table, apparently not overawed by the task of a Parisian champagne bottler convincing the greatest scientific and technical minds in France that he was about to save the country. He reached into his satchel and withdrew two large champagne bottles. "This, gentlemen," he said dramatically as he placed the bottles on the table, "will make the army of France the greatest army in all the world."

The dozen men around the table stared silently at Appert for a moment, hardly able to believe what they had just heard. "Do we presume correctly, Citizen Appert," one of them finally said in heavy sarcasm, "that you propose to defeat the enemies of France by throwing champagne bottles at them?"

Appert was unfazed. "Not at all, gentlemen. It is not the bottle; it is what is *inside* the bottle." He pointed to several green vegetables floating in some kind of liquid in the bottles. "These vegetables were sealed inside approximately three weeks ago, and as you can perceive, gentlemen, they are today just as green and fresh as the day I personally put them there."

As Appert began a long explanation of how he had discovered a way to keep food fresh by boiling it and then sealing it inside an airtight bottle, the society members seemed puzzled. For one thing, they were having difficulty understanding how the process was of any possible help to the French army. That was the very purpose of the society when it was set up earlier that year of 1800 by Napoleon Bonaparte; staffed by the best scientific and technical minds in France, its mandate was to find any scientific or industrial process that would help the French military. To aid that process, Napoleon was offering a cash reward of 12,000 francs (about $50,000 in today's currency) for a process that would prove useful.

To the society's dismay, the lure of the cash reward had attracted every nut in France. For months the society listened to a steady parade of crackpot inventors, psychics, astrologers, and spiritualists as they unfolded assorted ideas for miracle weapons—a magic dust that would put entire brigades of British soldiers to sleep, wooden wings that would enable French soldiers to become airborne, fused bombs attached to dogs that would be trained to run toward the enemy and blow themselves up. Now here was this champagne bottler, babbling on about preserving vegetables in bottles.

Nevertheless, they listened, hoping to find a glimmer of something useful. They were desperate: no one from France's scientific and technical establishment had arrived at any bright ideas yet, and an increasingly impatient Napoleon was wondering aloud why no French mind seemed able to come up with something, *anything*, to help the war effort.

But there was one society member, a military engineer, who had come fully alert during Appert's presentation, aware that the humble bottler was absolutely right: he had indeed come up with something of inestimable value to the French army—and something that would revolutionize war. He jotted down a few notes, then rushed off to Napoleon himself to deliver the news. Napoleon got it instantly, and with a few pen strokes, awarded Appert the prize of 12,000 francs, then commissioned him to build a factory to turn out many thousands of what Appert called his "food in a bottle."

And with those bottles, Napoleon's army went on the march, moving hundreds and thousands of miles in lightning campaigns. Freed from the necessity to forage or wait for slow, horse-drawn supply trains, La Grande Armée operated at distances no one had ever thought conceivable, since individual soldiers could now carry their food with them. This unprecedented mobility led the British to wonder how Napoleon was able to achieve it, given the limitations of early-nineteenth-century logistics. Soon the capture of a few

French soldiers who were carrying bottles led the British to the secret, and the British army's Ordnance Department issued an urgent plea to England's scientific and technical establishment: find out how the French process works, and invent an improved version.

If it seems improbable that a champagne bottler would come up with the key idea that made mass armies mobile, the next step was equally unlikely: The man who figured out how the process worked and how Britain could match it was not a scientist but an ordinary London mechanic named Peter Durand. Intrigued by the problem (and hopeful of turning a tidy profit if he could solve it), Durand carried out a number of experiments and realized Appert had hit upon his food preservation idea by basing it on his experience keeping the bubbles in champagne—an airtight cork seal kept in place by a tight wire basket. From there, he apparently wondered that if champagne bubbles could be preserved, why not food? The resulting process consisted of immersing fresh food in water, boiling it at high temperature to prevent putrefaction, then sealing it inside a bottle by means of an airtight cork (which, Durand discovered, was achieved by a mixture of various substances, including an especially pungent type of French cheese). An ingenious system, Durand concluded, but he also saw its weakness: glass. Even the thickest bottles were subject to breakage, not the ideal material for a combat environment. Durand had been working with tin, which led him to the idea of improving Appert's idea by sealing food in what he called "tin canisters" (later shortened to the term commonly in use today, "can").

Durand's innovation made the British army as mobile as Napoleon's, and created a subsidiary revolution in food processing, but its real significance was in the way it illustrated how the pace of scientific and technical change was beginning to speed up. Appert had arrived at his food preservation idea for armies on the march just a few months after announcement of Napoleon's prize, and Durand had devised an even better system less than a year after the British army's Ordnance Department had issued an urgent plea for someone to surpass it. The accelerated pace largely stemmed from political circumstances. Napoleon, the very apotheosis of total war, was a technically trained artillery officer who believed science was a critical factor in defining an army's cutting edge. As his creation of the Society for the Encouragement of French Industry demonstrated, he understood that this edge could only be kept sharp with a constant infusion of new scientific and technical developments. On the other side, the danger represented by a military genius equipped with the most up-to-date technology—the kind of genius who in an afternoon could smash three separate armies moving against his own greatly outnumbered force—mandated a constant drive to develop new weapons and technologies to keep him in check.

Napoleon's devotion to scientific advance (or at least the science that had important military applications) represented something of a sharp turn in direction for France. The peasants and small shopkeepers who created the French Revolution had little use for science, which they tended to regard as just another tool of the hated upper classes to oppress them. Thus, when Antoine Lavoisier—the great French chemist whose studies of the behavior of gases created modern chemistry—came before the revolutionary tribunal in 1792 on charges of being an aristocrat (and a tax collector, to boot), he was instantly condemned to death. An alarmed group of scientists came before the tribunal to argue for Lavoisier's life, citing his immense contributions to the French military—his improvement in the quality of gunpowder; his development of mass production techniques for it; and, supremely, his use of oxygen to construct better blast furnaces, which produced improved gunmetal. The scientists were careful not to cite any of Lavoisier's immense contributions to the science of chemistry, reasoning, correctly, that the largely illiterate revolutionaries wouldn't understand them. But in any event, the scientists' appeal failed to sway the tribunal, which concluded that all those experiments he conducted at his home laboratory were actually designed to suck all the air out of Paris and kill the entire population. "The revolution has no need of *savants*," the head of the tribunal sneered as Lavoisier was led off to the guillotine, there to join Louis XVI, among many others, in a bloody harvest of severed heads.

But only a year later, an agitated Gaspar Monge, the chief military engineer of the new revolutionary army, came before the Committee of Public Safety to warn that the tribunal's beheading of men like Lavoisier—and the general antagonism toward science—was putting the new republic in grave danger. The republic, he argued, was surrounded by hostile powers intent on suffocating the new nation in its crib; fighting had already broken out. Unless France put science back in the forefront, he warned, and organized all science in a government-run operation to develop new weapons, there was no hope of surviving the threats from without. The committee agreed, and in short order even scientists of aristocratic background were welcomed into the revolutionary fold. And just in time, too: as Monge predicted, the crowned heads of Europe, infuriated at the regicide of Louis XVI and determined to extinguish the revolutionary flame before it set the Continent afire, had sent their armies to invade France. The invasion was eventually beaten back by the surprisingly professional performance of the ragtag French revolutionary army, aided in no small part by a new weapon: a lightweight but powerful piece of field artillery that featured an adjustable hairline sight, improved gunpowder, and a variety of ammunition the gunners could use in different tactical situations, including deadly canister shot (basically a wire basket filled with steel shards that could slaughter entire ranks of infantry at 400 yards).

During this time a young artillery officer named Napoleon Bonaparte began his rise to military greatness. A master of those new mobile cannons, he demonstrated his remarkable abilities on the battlefield in 1798 when he led a French army on an expedition to Egypt. The Egyptian army, at least a century behind the Europeans technologically, foolishly attacked Napoleon, who blasted them to pieces, killing four thousand of them in less than forty minutes. Five years later, Napoleon took on a much more formidable foe: the Prussian army, ranked as the best in the world. To the astonishment of the European military establishment, in only thirty-three days a campaign of unprecedented speed and violence totally shattered the Prussian army, including Napoleon's capture of one hundred and fifty thousand prisoners. The fabled army of Frederick the Great, which had taken 140 years to build to peak efficiency, virtually disappeared. Total war had come to Europe.

Napoleon's chief weapon was his marvelous field artillery, whose mobility and quick rate of fire in the hands of a military leader who knew how to use it to best effect represented a dominant battlefield weapon. But Napoleon realized it was only a matter of time before his enemies—which included most of Europe—would develop their own dominant weapons. Napoleon took efforts to ensure that France remained in the forefront of the certain scientific-technical race to come. He created the École Polytechnique, France's MIT, where artillery officers and military engineers were trained in the latest scientific and technical advances; organized France's chemists into a government-run entity to work on perfection of gunpowder (they developed a process to make calcium nitrate, a key gunpowder ingredient, through the oxidation of ammonia); enlisted astronomers for the government payroll to develop new methods of sea navigation; and founded the Society for the Encouragement of French Industry, which led to Nicholas Appert's food preservation system. Napoleon had also taken a group of scientists to Egypt, where, among other things, he ordered them to study the idea of digging a canal from the Mediterranean to the Red Sea. (They concluded that it wasn't feasible, given the 30-foot difference between the water levels of the two seas.) While they were tromping around Egypt, the scientists discovered the Rosetta stone, which enabled translation of ancient Egyptian hieroglyphics, creating a sensation that led to an archaeological fever: Egyptology.

Yet for all his devotion to applied science, when Napoleon was presented with two ultimate weapons that would have made him master of Europe, he failed to see their potential. These failures proved the problems inherent in a system where a dictator, even one as talented as Napoleon, made all the decisions. Napoleon was scientifically literate, but he was not a scientist. And given that fact, he simply was not qualified to judge all scientific questions, which should have been left to a scientific committee. Almost certainly they would have provided a much more receptive atmosphere to the Montgolfier brothers and an intense American inventor named Robert Fulton.

• • •

In 1800 one of Napoleon's leading scientific lights, Jacques Conte, who had been among the scientific delegation that went to Egypt, came to him with what he thought was sensational news: two brothers in Paris had achieved the dream of manned flight. Well, not flight, exactly; manned ascent would be the more accurate term. Étienne and Joseph Montgolfier were papermakers who in 1781 felt inspired by the announcement of an award of 10,000 francs (about $42,000 in today's currency) by the French government to anyone who came up with a way to break the British siege of Gibraltar (then part of Spain, a French ally). The Montgolfiers had an interesting idea: a balloon, made of paper and canvas, whose interior gas would be expanded over an open flame; the expanded gas would lift the balloon into the air. Such balloons, the Montgolfiers thought, could sail over the British lines and bring supplies to the besieged Spanish, who were in the process of being starved out.

The Montgolfiers began a series of flight experiments, and on November 15, 1782, a 40-cubic-foot balloon of paper-covered canvas, carrying 400 pounds of ballast, rose to an altitude of 6,000 feet. By 1783 they had developed a balloon that traveled almost 8 miles at an altitude of 3,000 feet. Further tests involved carrying roosters, sheep, and ducks, the prelude to manned flight. By that time the Spanish had surrendered Gibraltar, and the immediate military reason for the Montgolfiers' balloon had disappeared. They nevertheless continued working on the concept, raising funds by charging Parisians for the ride of their lives.

But in 1800, when Jacques Conte first saw the Montgolfier balloons, he instantly grasped their military significance. Such balloons, he told Napoleon, would be perfect for reconnaissance: the French army could float them over the battlefield and behind enemy lines, with trained *aéronautes,* as Conte called them, aboard to observe with telescopes and spot enemy dispositions, routes of march, and other such vital military intelligence. And, Conte argued, there was nothing to stop the *aéronautes* from dropping bombs on enemy troops. From there, Conte proposed a dizzying vision of fleets of giant balloons dropping bombs on enemy capitals, destroying enemy armies, and serving as spotters for artillery.

To Conte's surprise, Napoleon did not seem especially enthusiastic. He reluctantly approved Conte's proposal for the founding of the government-funded Aérostatic Institute near Paris, which was to develop balloons into an ultimate weapon and train a new breed of warrior, the *aéronaute.* But only two years later, he ordered the operation defunded, telling Conte he saw no future in manned balloons as an instrument of war.

He was similarly unenthusiastic that same year when an American inventor named Robert Fulton came to Paris with an even more incredible idea than balloons with bombs. In a meeting with Napoleon and his top scientists, Fulton unrolled diagrams and drawings for a steam-powered warship and a strange vessel he called a "submarine," a warship that would sail underwater and attack surface ships. Napoleon was openly skeptical; it was scientifically impossible to power a wooden ship by building a fire belowdecks. As for a ship that sailed underwater, also impossible: How could the sailors aboard such a ship even steer? And he didn't see much of a future for still another Fulton scheme, something its inventor called an "unpropelled, contact-activated, zero buoyancy charge" (known today by its later name, torpedo).

Despite his misgivings, Napoleon reluctantly approved an expenditure of 40,000 francs (about $167,000 in today's money) to support a research effort by Fulton in France, during which time he would build prototypes of his miracle weapons and test them in French waters. In part, Napoleon was moved by Fulton's intense passion; a bitter Anglophobe, Fulton said his life's mission was to bring England to its knees, which is why, he said, he was offering his services to France (while omitting to mention that he could not find any enthusiasm for them in the American government). Only a year later, however, despite a promising test of Fulton's submarine in the Seine (it stayed submerged for four hours) and a successful test of his torpedo, Napoleon cut off all further development funds, insisting that Fulton's concepts would never work in actual combat.

With those two actions, Napoleon canceled two technologies that assuredly would have made him the master of Europe. It was a colossal error that two other nations would not make, at a terrible cost to the world—and themselves.

• • •

No nation suffered more from Napoleon's military power than Prussia, which by 1806 lay prostrate, utterly defeated. The Prussians, a hardheaded and practical people, immediately set about rebuilding their once-great power, a process that began with some hard thinking. They concluded that the army that had sallied forth at Jena and Auerstadt to be crushed in only a few hours lost because, simply, they were behind scientifically and technically. Prussia then began a huge bootstrap operation to vault the small nation back into the first rank of European powers, but this time as the most scientifically advanced military power on the Continent.

That effort began in the army, which was cleaned out of deadwood and its entire top leadership replaced. The Prussian Kriegsakademie was formed, a higher school of instruction for future officers that featured rigorous technical and scientific training. These officers, new regulations decreed, must be scientifically and technically literate, able to integrate the latest scientific and technical developments into the conduct of war. All these officers had one central, guiding principle drilled into their heads: war is a science, to be fought scientifically with all the latest tools scientists could provide.

And Prussia would spare no effort or expense, the government decreed, to create a first-class scientific elite. In 1810, just four years after the humiliation of Napoleon's triumphant march through Berlin, the government built the University of Berlin, intended to be the biggest and most scientifically advanced center of higher learning in Europe. For any budding scientist who thought his free tuition (courtesy of the state) would allow him to indulge in pure science, he was disabused of the notion by the university's first chancellor, who informed incoming students that "the state must replace with intellectual prowess what it has lost in physical strength." In other words, science in post-Napoleonic Prussia was to be applied science strictly for the improvement of the state. Obviously that meant science with military implications.

The University of Berlin was the crown jewel in a massive educational reform project that built a network of universities and technical schools. In an innovation, the system, heavily tilted toward science, enlisted prominent scientists who were not only to teach new generations of scientists but also continue their own researches in state-of-the-art university laboratories. The research university was born, with all the paraphernalia that later became familiar in academia: graduate-level seminars, colloquia, and specialized research centers inside the universities for advanced scientific research. The system also featured Ph.D. degrees as the required academic credential for scientific careers and the use of formal textbooks for instruction.

This system, which became the later model for scientific education in industrialized nations, was adopted whole when Prussia became part of a unified Germany in 1870. By that time it had vaulted Germany to world leadership in science and technology, largely aided by a subsidiary system—the *Technische Hochschulen* (higher technical schools), which trained hundreds of thousands of students in applied science and technology. They formed the operating arm

of the scientific advances that were pouring out of German universities, making practical the forward strides in scientific fields the
Germans came to dominate: chemistry, electrotechnology, and precision optics. These were the immense advantages that a Prussian
army first brought to the battlefield to destroy the Austrian army in
1866, and, to the shock of Europe, a German army—armed with new
quick-firing artillery and superior rifles—used to totally defeat
France in 1870 in only a few months.

Heady stuff, and the Germans, who by the beginning of the twentieth century had become preeminent in science, began to believe
that superiority would provide them with the edge to build a great
empire. "It brings army, navy, money, and power," the German
writer Friedrich Naumann declared in trumpeting the triumphs of
German science. The leaders of Germany agreed, and began to create a sprawling, government-funded scientific edifice to coordinate
and direct all scientific research in Germany—the clear implication
being that the state would control all scientific research. And given
their conviction that Germany's "day in the sun" had arrived, it did
not require much imagination to wonder what kind of science they
wanted. That was made clear in 1911, when Kaiser Wilhelm personally supervised the formation and construction of the Kaiser Wilhelm Gesellschaft, an umbrella organization assigned the task of
supervising and coordinating the work of all science research institutes in Germany. "Science is an element of national power," the
kaiser said in dedicating the building, and as a militarist, there could
not have been any question about his definition of "science." Still,
scientists who accepted their kaiser's belief that science existed for
the sole purpose of making a greater and more powerful Germany
found that life could be comfortable: they were given an honored
status in society ("*Herr Doktor,*" the highly polite and respectful
form of addressing a university professor, remains in use to this day),
paid very well, provided with the best laboratories and plenty of
time for research, and awarded lifetime tenure as a university professor. Of course, this kind of status and privilege was reserved only
for the scientists who shared their government's view that science
existed solely to promote the "national interest."

The German scientific renaissance came at the expense of Britain,
which by 1860 was at the apex of its power: Ruler of 12 million

square miles of territory (25 percent of the planet's land surface); although it represented only 2 percent of the world's population, it turned out 40 percent of the world's industrial production. Its preeminence in science and technology was unquestioned. But Britain, convinced this mastery would never end, became complacent, and its lead in science began to erode. Without a major war to stimulate scientific advance, British science started to fall behind Germany and the rising industrial might of the United States. For example, British scientists discovered the properties of the mysterious force called electricity, but neither government nor industry evinced any interest; as a result, the leadership of what would prove to be a crucial scientific advance for modern industrialism passed to Germany and the United States (where George Westinghouse was astounded to discover that the British government made no effort to utilize the discovery of alternating current by the brilliant physicist Michael Faraday, a lapse that allowed Westinghouse to develop it). By 1890, Germany's share of world industrial manufacturing surpassed Britain's. In 1902, the steel produced by just one American manufacturer, Andrew Carnegie, surpassed the entire output of Great Britain. There were no British experts at his plants, seeking to learn the latest American steel manufacturing techniques, but there were polite men from the East who were busy paying license fees for the rights to use Carnegie's techniques while they collected every technical paper they could find. Carnegie happily took their money, all the while wondering why these men from a small country would want to make so much steel. He did not have to wait long for the answer.

• • •

Like Prussia, Japan underwent a traumatic event that would stimulate its own scientific renaissance. It was not a military defeat, such as the disaster inflicted by Napoleon upon Prussia, but something that to the insular Japanese was even more shocking: one day in the spring of 1854, four U.S. Navy warships commanded by Commodore Matthew C. Perry sailed into Tokyo Bay on what Perry described as a "visit." It was nothing of the kind; in fact Perry was playing gunboat diplomacy, arriving in Japan uninvited with a stark reminder of the superiority of Western science and technology, as exemplified by the big warships with their bristling cannons—all the better to get Japan to sign favorable treaties granting a monopoly to American and European traders.

The Japanese realized at once how far behind they were scientifically. They still lived in a quasi-feudal society, where the code of the *samurai* prevailed; the grip of religious orthodoxy was strong; and science had barely advanced beyond the metallurgy necessary to produce strong swords. The shock of Perry's visit led to the decision by Emperor Meiji ("the Enlightened One") to make Japan a modern nation virtually overnight. First he created the Iwakura Mission, which sent representatives all over the world to examine science and technology—the keys to Western domination, the mission concluded. The mission's members went on to tell the emperor that unless Japan became at least the scientific equal of Western nations and built a modern military, it would soon go the way of China: an insular nation whose failure to develop science and technology led to weakness, the kind of impotence that was allowing the Western powers to carve up the nation like a turkey. Japan would suffer the same fate, the emperor was warned, without a major effort by Japan to leapfrog centuries of insularity and disinterest in science.

The emperor agreed, and initiated what still ranks as the most dramatically rapid transformation of a society in all history. He sent thousands of promising students to study abroad (75 percent of them enrolled in science courses), and enlisted hundreds of translators in a crash program to translate every Western science text that could be obtained into Japanese. Japan's representatives abroad were ordered to collect and send back home every scrap of scientific information they encountered. The best talent from the Prussian Kriegsakademie was hired to overhaul and modernize the Japanese military structure. He also hired battalions of foreign science teachers and lured them to Japan with very high salaries to teach science in Japanese schools. In 1894 the foreign teachers, no longer needed, were sent home; incredibly, in only forty years, Japan had become scientifically literate and now had a first-class military armed with the most modern weapons. That year, the rest of the world suddenly became aware of this remarkable transformation when the Japanese pounced on China, achieving a total victory that gained them Formosa (now Taiwan) and Korea. And eleven years later, there was an even greater shock: The Japanese took on a major Western power, Russia, in a war that featured the naval battle of Tsushima, when a Japanese fleet destroyed twenty-six of twenty-nine Russian warships. Significantly, the Japanese victory was achieved by superior technology, including long-range naval guns and the first use of radio in naval warfare. Those radios were aboard Japanese ships, giving their commander a huge advantage; he could deploy and maneuver his ships rapidly while his radioless Russian counterpart, relying on the traditional flag signaling system (which was slow and could be obscured by bad weather or the smoke of battle), found it impossible to react quickly to the maneuvers of the Japanese fleet.

• • •

The sounds of battle at Tsushima echoed around the world like the clanging of an alarm bell. David had slain Goliath, the world's largest military power, an achievement due almost exclusively to Japan's mastery of science and technology. One of the world's largest navies had been humbled by a smaller navy equipped with superior technology, and Japanese armies, armed with the most advanced rapid-firing artillery, had defeated the "Russian steamroller," the great infantry mass that had torn apart Napoleon's army and had ever since dominated the Eurasian landmass.

The shocking Japanese victory provided further proof, if any was needed, that science and technology determined success in war. Like the Japanese, the Russians many years before had bootstrapped themselves into a world power. In the eighteenth century, Czar Peter the Great realized Russia had fallen behind scientifically and technologically, and conducted a crash program to catch up. To that end, he set up the St. Petersburg Academy, a government-funded scientific research center, paid some of Europe's top scientists to work there, and applied them to the task of training a whole generation of Russian scientists and updating Russia's creaky military machine. Peter himself often toured Western Europe in disguise, visiting naval shipyards and military installations to observe the latest developments in weaponry. What he could not buy he obtained by theft; a secret police force was assigned the job of stealing blueprints, plans, technical specifications, and formulas for technology, to be duplicated in Russia. But after Peter's death, Russia was racked by internal divisions and political unrest; gradually the huge military machine Peter built began to rust. By the time of the Russo-Japanese War in 1905, it had fallen into serious disrepair.

The Russian disaster in the Far East took place at the high point of the greatest arms race in history, a race that was fundamentally a scientific-technical competition to develop bigger and greater weapons, a relentless (and by now institutionalized) drive to find the magical silver bullet that would instantly confer military invulnerability or superpower status. The race had been set off by the age of Napoleon, the advent of true total war. The scale of Napoleonic war was unprecedented: three hundred thousand men locked in mortal combat at Austerlitz, a million troops fighting in Russia, more than six hundred thousand at Waterloo. Entire populations had become swept up in war—as in Spain, where civilians were forced to choose between the French and British forces occupying their homeland. Those Spaniards who chose to fight against the French (and contributed the word *guerrilla* to the lexicon of war) found themselves in a war of no quarter where an unarmed civilian was just as likely to be shot as an armed soldier.

In 1816, as the Napoleonic era was drawing to a close, the novel *Frankenstein* was published. Above all, the book was intended as a clear warning about the evils of science run amok, but no one appeared to be listening. Science, now harnessed more firmly than ever to the needs of the state, was busy providing ever greater tools of war—bigger guns, more powerful gunpowder, mightier warships, deadlier infantry weapons. Any political power that could afford the cost was building mass armies and huge fleets, mostly on the perception that if they didn't make the investment, a rival power would, with possible consequences that were unthinkable. Like the Red Queen in *Alice in Wonderland,* major powers found that they had to run furiously merely to stay in place.

Most of the scientific-technical effort was directed at improving explosive power, a furious race that occupied the attention of chemists in half a dozen countries. The first great scientific breakthrough came in 1846, when Swiss chemists discovered that combining cotton wool, sulfuric acid, and nitric acid produced an explosive of tremendous power. Called "gun cotton," it was three times more powerful than conventional gunpowder and had the added advantage of being smokeless and flashless, perfect for artillery. Cannoneers had long complained about the difficulty of find-ing targets on a battlefield obscured by the clouds of heavy white smoke generated by gunpowder weapons, the chief reason why armies were clad in bright uniforms—so opposing sides could be dif-ferentiated in all that smoke. Twelve years later, Italian chemists developed an even more powerful explosive, a liquid form called nitroglycerin. The problem was its high volatility, a disadvantage solved by a Swedish armsmaker and chemist named Alfred Nobel, who discovered that combining nitroglycerin with a porous mineral known as *kieselguhr* (or, more popularly, diatomaceous earth) into a compact solid and attaching a safe and reliable blasting cap created an explosive he called dynamite.

The rapid development of bigger and better explosives ran in tan-dem with another scientific race, a metallurgical one to develop metals capable of withstanding the new explosive power. In 1850 Napoleon III of France offered a cash prize of 20,000 francs ($50,000 in today's money) for a cheap process that would produce new steel capable of standing up to the growing power of guns. The prize was collected by an English metallurgist, Henry Bessemer, who found that blowing air through molten metal during the production process produced a superhard steel that would withstand even the most powerful cannonballs. The Bessemer process set off still an-other arms race as metallurgists, chemists, and physicists began a cycle of more powerful explosives, better steel, more powerful ex-plosives, better steel—a hideously expensive competition whose subsidiary effects created modern industrial civilization. But its im-petus was strictly military; the German steelmaker Alfred Krupp built locomotives and steel rails for the new innovation of railroads, but his steel, considered the finest in the world, grew out of his government-funded efforts to develop the world's most advanced artillery for what he called "Germany's day in the sun." At the same time, Europe's other great industrial power, Great Britain, was busy trying to find new scientific breakthroughs that would keep the fighting edge on its greatest (and essential) weapon, the Royal Navy. In the 1870s British physicists and ordnance experts arrived at a crucial insight that suddenly rendered every steel warship in the world obsolete. The British scientists realized that trying to punch through steel with round cannon shot, no matter how powerful, ultimately would be defeated by thicker and better steel. The solution, they concluded, was ammunition that actually was a two-stage explosive: something to punch through steel, then an explosive that would go off on the other side. Reaching back to the English longbow arrow design, they came up with a shell that had a cone-shaped warhead that ended in a pointed nose. The nose, of superhard steel, would punch through steel, by force of the impetus provided by high explosives in the gun; then the explosives in the shell would go off. The focus was naval: a shell that could cut through the steel armor of a ship's decks and explode in the ship's vitals.

The British had devoted much study to the militarily epochal event of March 9, 1862, at Hampton Roads, Virginia, when an all-iron Union ship, the *Monitor*, fought an inconclusive artillery duel with a Confederate counterpart, the *Merrimac*. The encounter was militarily a draw because both ships fired cannonballs at each other, ammunition that bounced off their thick steel hides. Nevertheless, it was clear that the moment both ships appeared, naval warfare had changed forever; the days of the wooden warship were over. The *Monitor* contained an even more important technological development, one not so obvious as its revolving turret; it was a radical new design, the screw propeller. This innovation, borrowed from Archimedes' screw device that lifted water from the Nile to irrigate fields in Egypt, was the brainchild of John Ericsson, an émigré Swedish engineer who convinced Abraham Lincoln that he could build a new kind of warship capable of sailing up Confederate-controlled rivers, impervious to the fire of shore batteries. Ericsson's innovation meant that all-steel ships could move at high speed on any navigable waterway without being subject to the vagaries of winds while armed with guns in revolving turrets—another important technical innovation that allowed warships to fire from any position at sea.

The result was still another accelerated arms race between steel and shell: faster, sleeker warships with thicker steel; new artillery shells to penetrate the armor; still thicker steel; new artillery shells to penetrate that steel, and on and on. The race finally reached an apex in 1907 when the Royal Navy unveiled what was universally regarded as the ultimate naval weapon: the battleship HMS *Dreadnought*, a 15,000-ton behemoth. The ship encapsulated the latest advances from a host of scientific disciplines—thick steel able to shrug off all but the biggest artillery shells, gyro-stabilized steering, rifled naval guns in revolving turrets capable of firing shells up to nearly 8 miles in range, centralized fire control using telegraphy, and huge propellers to push all that steel at speeds of up to 20 knots per hour.

No one could conceive of a superior weapon to the *Dreadnought* except a bigger and better *Dreadnought*—which is why a naval race then broke out as nations rushed to construct fleets of Dreadnought-class superships. Among the few dissenting voices to this mad battleship race was an unlikely one, an Irish science teacher and part-time amateur scientist named John P. Holland. He was convinced that building bigger battleships to checkmate the *Dreadnought* was a hopeless cul-de-sac, since the British would simply build even bigger and better battleships, as their science and technology demonstrably were capable of doing. Holland had been thinking about this matter since 1873 because he was a fanatic Irish patriot and member of the Fenians, an Irish revolutionary movement. He believed the key to loosening the British grip on Ireland was to cripple Britain's sea power—and that meant defeating her capital ships. His researches on a possible countermeasure finally led him to Robert Fulton's concept of a submarine that had failed to impress Napoleon. Holland instantly saw the solution he was looking for; if he could perfect that idea into a true submersible warship capable of attacking, and sinking, a battleship at a distance, the Royal Navy was doomed.

A very ambitious goal for a schoolteacher, but Holland got the Fenians to underwrite a research and development effort in which Holland actually built and tested a small submarine. In the process, by 1890—at which point he had immigrated to the United States to become an American citizen—he had devised a number of technological breakthroughs that made the submarine a true warship: a periscope for observation; ballast tanks to allow the boat to dive and surface; hydroplanes to make it navigable underwater; new torpedoes that could be launched with a compressed air motor; and, most ingenious of all, a double propulsion system (a gasoline engine for travel on the surface; another, powered by batteries, for underwater travel). It was a remarkable technological achievement, and on May 17, 1897, Holland took his perfected design to a New Jersey shipyard for a public demonstration, hoping to attract government funding.

With Holland himself at the helm, the small submarine spent four hours diving and surfacing and, for good measure, fired a small, warheadless torpedo at a surface target, hitting it squarely at a range of 300 yards. A *New York Times* reporter on the scene, not knowing quite what to make of it all, cautiously wrote a dispatch in which he said Holland's submarine "may nor may not play an important part in navies of the world in the years to come" (which covered all the possibilities). The U.S. Navy and the Royal Navy seemed only mildly interested in the demonstration, but several observers from the German Kriegsmarine were paying closer attention; shortly thereafter, they bought the German licenses for Holland's patent. The reaction of Clara Barton, founder of the Red Cross, was quite different: When she read the *Times* dispatch, she immediately rebuked Holland in a letter that accused him of inventing "a dreadful weapon of war." Holland replied that his submarine would become the ultimate naval weapon, one so powerful as to serve as a deterrent to international conflict. Holland would not be the first to become subject to this delusion.

Tragically, Barton turned out to be right. On September 21, 1914, the German submarine *U-9*, an enlarged version of the undersea boat Holland had unveiled some seventeen years before, in a matter of minutes sank three British cruisers with three torpedoes, killing fourteen hundred sailors. For a few moments at least, Holland achieved his dream of the Royal Navy brought low by his submarine—small comfort, perhaps, considering the destruction his invention would wreak in the years ahead. Only some of the victims would be British.

• • •

While the steel-versus-shell race was under way at sea, a similar race was being enacted on land as nations exerted their best efforts to develop bigger and more powerful artillery along with more powerful infantry weapons. The pace of technical development was astonishing: by 1862, the Union armies in the American Civil War were using a huge rifled cannon known as the Parrott Gun, which had the then remarkable range of more than 6 miles. Within two decades there were breech-loading artillery pieces, explosive shells to replace cannonballs, and the ultimate development in artillery, a recoilless gun. That gun, the revolutionary "French 75," was a seventy-five-millimeter, breech-loading weapon with a hydraulic system that absorbed the recoil and kept the gun stationary. That allowed for a high rate of fire, since the gunners no longer had to wait for the recoil, then reposition the gun. By 1906, trained French gunners could get off up to twenty rounds a minute, which meant that massed batteries of such weapons could deliver devastating barrages of fire on entire sectors of a battlefield.

Infantry weapons also had become much more deadly, especially the weapon that enabled European powers to maintain control over huge colonial territories with very few troops: the machine gun. Brought to perfection in the 1870s by the British engineer and inventor Hiram Maxim, the weapon represented a fusion of several scientific disciplines in one very efficient killing machine—a system that used the explosive gases set off by one bullet to propel the next one automatically; a belt-fed ammunition feed system mathematically calculated to feed bullets into the firing chamber at precisely the right moment; and a water-filled steel jacket around the barrel that kept the temperature consistent and prevented the barrel from melting or cracking from the intense heat generated by explosive gases. No better man-killer had ever been seen in war, a scientific and technical triumph that would become responsible for the deaths of more men in war than any other weapon.

For the people of less scientifically advanced civilizations, encountering the machine gun for the first time was truly terrifying. In 1898, during a Mahdist uprising against British colonial rule in Sudan, nearly fourteen thousand Dervish warriors came upon some five hundred British soldiers drawn up in a classic infantry "square" in the middle of the desert. Superb light infantrymen, the Dervishes immediately deployed for attack against what seemed to be a target they could easily overwhelm. They didn't recognize the strange-looking devices the British troops had with them—in fact, six Maxim machine guns. The standard Dervish tactic was to rush at high speed toward an enemy, yelling blood-curdling war cries, then falling upon them with razor-sharp curved swords that could sever a man's head with just a flick of the wrist. The Dervishes rushed toward the British square, and when they had come within 300 yards, the British guns opened up. Within forty minutes more than eleven thousand Dervishes lay dead in heaps; the rest fled in terror.

After the battle—if a slaughter can be so termed—the commander of the British forces noted, "The battle was won by a quiet scientific gentleman living in Kent." Just so; Hiram Maxim's invention had enabled a mere handful of European soldiers to completely dominate an entire nation and its people. As the writer Hillaire Belloc summarized it in his poem "The Modern Traveler":

> Whatever happens, we have got
> The Maxim gun, and they have not.

The machine gun and a host of other scientific and technical wonders would mean the end for a number of ancient civilizations as they fell before the weight of a European civilization armed with advantages native peoples could not hope to overcome. As the Incas who hurled themselves against the Spaniards and the Dervishes who did the same against the British proved, there was no sense in fighting the Europeans on their own terms. That was simply a guarantee of defeat.

The colonial subjects came to believe that the Europeans were supremely invincible. They could not have known that these creatures, with their terrifying science and technology, were about to do the one thing their subjects could never have predicted they would do.

The Europeans were about to commit suicide.

• • •

Very few saw it coming. There were many, like the submarine inventor John P. Holland, who insisted that the sheer destructiveness of the modern weapons that science had provided would make war so terrible, no sane nation would fight one. What sane nation would jeopardize the great industrial civilization that had brought so much benefit to so many people? What sane nation would allow the cultural treasures, the great cities, and the flourishing towns of Europe to be destroyed in the name of war?

Yet, the signs were there. The mad race for an ultimate military security—to develop the most advanced and destructive armaments, to construct mighty military edifices, to build huge fleets, to conscript millions into vast armies, to overlay the Continent with a network of railroads to move troops rapidly, to stockpile mammoth amounts of ammunition and guns, to prepare war plans that called for armies of millions to fall upon each other in mass battles—amounted to a gargantuan powder keg, needing only a single political spark to set off Armageddon. For nearly a hundred years, Europe had enlarged the Napoleonic concept of total war with scientific and technical advances to make it even more total. Science finally had been mated indissolubly to war, creating a juggernaut that no one was able to stop. But politicians and generals thought that although modern total war would be incredibly violent, like nothing ever seen before, it would be short. European generals dreamed and planned for a short war in which the outcome would be decided in one great Napoleonic campaign, the decisive blow, using the full might of modern armaments, to destroy an enemy totally. No one who directed any of these mighty military machines seemed to doubt that the fruits of modern science—metallurgy, chemistry, industrial machinery, radio, turbines, diesel engines, hydraulics, fire-control mathematics, gun ballistics, optics—provided an irresistible force that guaranteed victory.

But when the great cataclysm of World War I finally arrived, events soon proved this belief a tragic chimera. By the end of 1914, the mass armies were locked in stalemate and, inevitably, they turned to the one weapon they believed would end that stalemate and bring victory: science. And that would cause the gravest crisis of conscience in the history of science, one that remains unresolved to this day.

It was a crisis summarized in the career of one of the century's greatest scientific minds, a mind that simultaneously could provide the science to improve mankind's existence and the science to destroy it.

• • •

In 1909, the German chemist Fritz Haber electrified his scientific discipline with an amazing discovery: recovering nitrogen from the atmosphere by "fixing" it in a process that involved synthesizing ammonia. The implications of that discovery were enormous: with the process, agricultural fertilizer could be mass-produced at low cost, the development that set off the agricultural revolution in which per-acre yields increased a hundredfold, allowing nations for the first time to produce enough food to feed their growing populations. Fertilizer had been relatively expensive up to that point, since its key ingredient, sodium nitrate, had to be imported, mostly from large deposits in Chile. Haber's process eliminated the need for that chemical in fertilizer production. The feat would earn him the Nobel Prize in chemistry, and in 1911 he was named head of one of the scientific institutes Kaiser Wilhelm had set up to establish German preeminence in science.

Haber was busily at work at the institute in 1914 when an alarmed German High Command approached him for help in solving a serious logistical problem, one that involved a subject Haber knew well: sodium nitrate. The problem, Haber was told, stemmed from the fact that sodium nitrate was a key ingredient in explosives. Germany's stock of the chemical would be exhausted by 1916, given the vast expenditure of artillery ammunition the war required. And since a British naval blockade had cut off Chilean supplies, Germany faced the prospect of running out of ammunition. Could *Herr Doktor* Haber find a solution?

In short order, Haber found the solution, which amounted, basically, to a refinement of the process that had revolutionized world agriculture, this time involving a new process of making gun cotton that eliminated the need for sodium nitrate. Haber had taken a fateful step, for he was now using his scientific genius not for the betterment of mankind—as he once claimed—but for its destruction. And there was worse to come. In 1915 the German High Command approached him again, this time with a problem involving battlefield tactics. Artillery, even great barrages of fire, was proving insufficient in driving enemy troops out of their trenches; they simply dug deeper. Was there some kind of chemical Haber could devise that would be fired into enemy trenches, forcing the troops out? Haber's solution was a relatively mild form of chlorine gas, highly irritating (though not fatal) to the human respiratory system. But the gas, launched toward enemy trenches by large blowers, proved subject to wind conditions, and was easily defeated by soldiers holding wet handkerchiefs over their noses and mouths. The trench deadlock continued, and desperate for something to end it, the High Command turned to Haber again. This time he had a drastic solution in mind: poison gas, the greatest evil ever spawned by the union of science and war until the advent of thermonuclear weapons.

In 1916 the German army created the Chemical Warfare Service and named Haber as its chief, with a mandate to come up with poison gases that could be packed inside the warheads of artillery shells, then exploded in enemy lines and spread deadly fumes capable of wiping out entire armies. Working nearly around the clock, Haber finally came up with phosgene, which could kill a man in seconds, and mustard gas, which even if not fatal, almost invariably caused blindness and severe lung damage. That year, these terrible new weapons were unleashed against a section of French-held front; several thousand soldiers were killed outright and many thousands of others ran in panic. The Germans didn't have the mobility to exploit the success, and the Allies quickly came up with countermeasures, especially the invention of the gas mask.

But poison gas would now forever be linked to the name of Fritz Haber, an odiousness that did not appear to bother him. He was puzzled when he heard about the controversy that broke out in Britain when a prominent British chemist, Frederick Soddy, refused to do any research connected with poison gas, and was threatened with incarceration in the Tower of London as a "traitor." To Haber, the Soddy case made no sense; he assumed that Soddy was as fervid a British patriot as Haber was a German one. And in a time of war, a scientist's first duty was to his country, not to mankind or pure science. If a scientist was called upon to invent new and greater ways of killing other human beings, that was an unfortunate but necessary by-product of war.

For that reason, Haber felt no discomfort when his closest friend, Albert Einstein, berated him for using his great scientific talents to slaughter so many of his fellow human beings. In Haber's view, Einstein was a hopeless dreamer when it came to politics, a brilliant scientific mind who did not understand that Germany's very existence hung in the balance. He insisted that in a total war, where science was the major determinant of national survival, a scientist was just as much a soldier as the man with a rifle on the front line. Despite the great chasm between them on this subject, Haber and the lonely pacifist remained good friends. The same, however, cannot be said of relations between Haber and his wife. Appalled that her husband was personally responsible for so many deaths, she begged him to stop and follow the example of his close friend Einstein, who refused to have anything to do with any science that involved war. She and her husband had increasingly bitter arguments on the subject; finally, when she realized she could not sway him, she committed suicide.

Even this tragic event in his life failed to change Haber's mind. At the end of the war, he was shocked to learn that the victorious Allies were considering bringing him up on war crimes charges. The plan was dropped after a number of scientists rushed to Haber's defense, notably the distinguished British biologist J.B.S. Haldane, who argued that the war was intrinsically evil before the invention of poison gas, and would have remained evil because of high-explosive shells, the tank, the submarine, and barbed wire. So how could Haber take something that was already evil beyond comprehension and make it even more evil?

An interesting question. The answer would come sooner and more terribly than Haldane could have realized.

7

The Sorcerer's Apprentices

The debacle of liberal science can be traced to the
moral schism of the modern world which so
tragically divides enlightened men.

— WALTER LIPPMANN

World War II was won by the Allies on August 18, 1940, when two British citizens, one of them carrying a small black trunk, arrived in Washington, D.C., aboard a plane that landed in a remote section of a military airfield. The two men, whose diplomatic passports described them vaguely as "members of the Foreign Office," were whisked to a room at the War Department. There, before an expectant audience of five Americans, they opened the trunk and withdrew several mechanical devices and three typewritten pages.

In that moment, Nazi Germany was doomed. Those devices and the three pieces of paper amounted to a synopsis of the scientific and technical edge that Great Britain held over its enemy. Now this Aladdin's Cave of scientific wonders would be mated to American science, technology, and industry, creating an unstoppable juggernaut and setting off the greatest scientific-technical revolution in history. Within five years it would win the war and transform the planet forever; its consequences (and its benefits) remain with us today.

The significance of what at first glance appeared to be an unremarkable event extended even farther. For one thing, it marked an unprecedented and extraordinary surrender of a nation's greatest science secrets; no nation in history had ever voluntarily revealed so much, even to an ally. For another, it marked an important transition in the history of science: from that moment, the leadership of science passed from the Old World to the New World, where it has remained.

• • •

One of the two Englishmen who journeyed to Washington was Henry Tizard, Prime Minister Winston Churchill's chief scientific adviser. His top-secret mission, assigned by Churchill personally, was to convey to the Americans the treasures of British science, the treasures that had kept Britain alive under the pounding of German bombers and imminent invasion. Churchill wanted the United States to join Britain in the war effort against Germany, and his decision to share his nation's greatest scientific secrets had a twofold purpose: One, Britain needed the tremendous American industrial capacity to provide the huge amount of armaments a hard-pressed British defense industry was having trouble producing; and two, the British, desperately short of money, needed an infusion of American science and technology to keep the cutting edge on their own science and technology. Essentially Churchill was trying to effect a marriage between the sciences of the two nations. (There was a third, more subtle, purpose: Churchill believed he could win President Franklin Roosevelt's favor more readily if he made a magnanimous gesture—and revealing a nation's entire range of scientific secrets certainly qualified.)

The small group of American scientists who got a first look at the treasures to emerge from the black trunk were astounded at the sheer range of the British scientific achievement the contents revealed: twenty-one technologies in eight different scientific disciplines. What the Americans saw suddenly lit up a number of dark corners that had been intriguing them for some time. For example, there was a cavity magneton, capable of generating intense, short wavelengths, the explanation for why the outnumbered British were winning the Battle of Britain: They could divine the exact size and heading of German bomber formations because those short wavelengths allowed the British radar system to pick up approaching German formations up to several hundred miles away, which meant the RAF could unerringly vector their fighter forces to those formations. With the fuel saved because fighters did not have to use precious gasoline to hunt their prey, British pilots could stay on station longer to attack the German bombers. Additionally, there was an ordinary-looking piece of rubber, which concealed its technical secret: British scientists had developed a special form of rubber capable of absorbing, then enveloping, an intruding bullet. Wrapped around the gas tank of a fighter, the rubber formed what was called a "self-sealing gas tank," which eliminated a pilot's greatest fear: exploding in a fireball when even a single bullet managed to penetrate the tank.

But the greatest treasure of all was the most prosaic: that three-page typewritten document. To the Americans' astonishment, it revealed that a select group of British physicists had begun work on something they code-named "Tube Alloys Project," a deliberately vague title meant to conceal what was in fact a project to build an atomic bomb. According to the memorandum—actually a progress report to the government on the feasibility of the project—the physicists had concluded that the concept of utilizing the tremendous power unleashed by the fission of certain uranium atoms for a bomb was scientifically feasible. However, they added, a considerable number of technological hurdles remained, and overcoming them would take a huge amount of money (cash the British did not have). But it could be done.

The effect on American science of what would become known as "the Tizard Mission" was nothing short of electric. More than a year later, one of America's most prominent scientific minds, Dr. Vannevar Bush, president of the Carnegie Institution and former dean of engineering at the Massachusetts Institute of Technology, was in President Roosevelt's office to propose an idea. Bush didn't waste much time telling FDR what he already knew—that Hitler was a serious threat to the future of the world, as was the military power of Japan in the East; that Germany had been preeminent in science and still retained leadership in a number of scientific fields with strong military implications; that an enlistment of German science with Nazi ideology had implications nearly unthinkable in their consequences; that a beleaguered Great Britain desperately needed American help; that a total Nazi domination of Europe would initiate a new Dark Ages and threaten the United States; and that there was an urgent need to set American science working to prepare for a war that was about to involve the United States.

Bush proposed creation of a new entity he called the National Defense Research Committee (later renamed the Office of Scientific

Research and Development). At root, the concept was simple: The organization would recruit the cream of American science in every scientific discipline. They would be assigned to work on specific military problems; any of those recruited who were of draft age would be officially exempt from the new Selective Service Act. More significantly, the organization would be free of military or political control. The money would be allocated to the organization, which would have total say on just how the money would be divided among the various research and development projects; scientists would decide scientific matters.

Somewhat to Bush's surprise, FDR immediately agreed, saying simply, "Okay, go do it," and with characteristic decisiveness named Bush as his first science adviser. While Roosevelt rounded up the necessary funds, Bush and three other leading science academics rented out a New York hotel room. There they invited forty of the nation's leading scientists and enlisted them, in greatest secrecy, for the new organization. Each scientist was to recruit a similar number of the top scientists in his field, a geometric progression that within months resulted in some thirty thousand of the nation's best scientists and engineers enlisted for war.

The new organization built on America's first experience with organizing science for war, during World War I, when President Wilson established the National Research Council, which functioned as a clearinghouse for scientific research projects connected to the war effort. By 1918 it was so busy that entire scientific research institutions were springing up; both MIT and the California Institute of Technology largely owe their existence to this effort. As the Americans discovered, modern war required specific, coordinated scientific efforts directed toward solving specific threats. Among the more dangerous were the German U-boats. Both Great Britain and the United States, because of their early lack of enthusiasm for John Holland's submarine weapon, found themselves totally unprepared when the German weapon was unleashed in 1914. A desperate British Admiralty staff, which had never seen a submarine, much less understood how it functioned, sat around trying to think of ways to defeat the German *Unterseebooten* fleet. Among the more dimwitted ideas this brainstorming produced was training sea lions to munch on periscopes, and enlisting blacksmiths who would be taken out to sea on picket boats to smash U-boat periscopes with hammers. An infuriated first sea lord Winston Churchill ordered an end to such nonsense and told the admirals to get the scientists cracking. In the summer of 1915 the admiralty created the Board on Invention and Research, an organization of physicists, mathematicians, and marine engineers assigned the task of defeating the U-boats. Within two years the scientists succeeded, in the process inventing the now familiar tools of antisubmarine warfare: depth charges; hydrophones (an acoustical detection system whose science would lead eventually to stereophonic sound systems); sonar; and the convoy system, the result of a counterintuitive deduction by mathematicians, who found that merchant ships packed closely together were actually safer than if they sailed alone.

The British Admiralty's early comic opera attempts to defeat the German U-boats with sea lions and blacksmiths marked the last time in that war that the Allied military had much of anything to say about how science would be used in war. From that point on, civilian scientists decided scientific matters. Conservative military opposition to the idea that civilian scientists would tell soldiers how to fight wars quickly melted away for the simple reason that the science worked. Both British and American admirals thought the idea of gathering merchant ships in convoys protected by escorts perfectly insane, but the scientists turned out to be right; losses to U-boats fell dramatically. Similarly, generals thought the idea of crossing a no-man's-land in a steel box atop a continuous chain of segmented tracks was crazy, but by 1918 the tank was fulfilling its promise; just as its designers predicted, it could cross muddy ground pockmarked with shellholes, smash its way through barbed wire, and crush enemy trenches. And the British Royal Flying Corps thought the mathematician F. A. Lindermann had lost his mind when he told its pilots he had arrived at an algebraic formula that showed they could get out of spinning nosedives without crashing. In the face of such skepticism, a determined Lindermann taught himself how to fly, then demonstrated that his formula worked. British pilots stopped dying from spinning nosedives.

More than ever before, World War I institutionalized scientific progress and technical innovation, in the process teaching warring nations several important lessons. One, if a little technology is good, then a lot of technology must be better; there is a scientific or a technological solution for every battlefield problem. Two, scientific resources are as precious—perhaps even more so—as battleships and armies, and should be protected. Only late in the war did the British discover that they had made a hideous mistake in not exempting from combat service scientists, technicians, and engineers. Too many of them were being killed in combat, a senseless waste that moved the eminent physicist Ernest Rutherford, head of the Cavendish Laboratory, to complain that his most promising protégé, a brilliant nuclear physicist named Henry Moseley, had been killed at Gallipoli. Consider, Rutherford bitterly noted, how Moseley would have served his country better at Cavendish than "at the end of a Turkish bullet."

As the war clouds gathered over Europe again in 1938, a group of prominent British scientists and economists met in their London club, the Tots and Quots, amid growing concern that a new war, building on the science and technology of the last, would pose an even greater threat to Britain. And like the last war, they concluded, Britain's real salvation rested in its ability to harness scientific and technical talent to maintain a superiority in both fields. Further, they recommended that science advisers—scientists with broad experience in a wide range of scientific disciplines—be appointed to the government to recommend where precious resources should be allocated. Their recommendations were outlined in a thin paperback book, *Science in War*, and would serve as a model for both the British and the American scientific war efforts.

What was decided at the Tots and Quots was not universally admired within the scientific community. There were a number of scientists who continued to believe that to use science in the pursuit of war could not be justified, and several recommended that scientists organize and refuse to do any war work. By their reasoning, if all the world's scientists refused to do war work, then wars would not be fought. Not an especially sophisticated analysis, considering the presumed intelligence of the people who made it, but sufficient to worry Solly Zuckerman, one of the scientific advisers to the government. "We cannot invest pure scientific knowledge with any inherent moral distinction," he said in defense of the militarization of British science as World War II began. "That is imparted in the way science is used."

That argument, a classic one made consistently by those who harness the needs of the state to science, bothered the distinguished

British scientist and novelist C. P. Snow, a man with a foot in the two camps of science and letters. It's all well and good, he argued, for scientists to decide that their science, however "pure," can be turned to the needs of the state in the name of patriotism, but science then was on a moral escalator, very difficult to dismount. How does a scientist know when it's time to get off? Why didn't Fritz Haber realize that at a certain point on his own escalator, it was time to leave when the state wanted him to provide a terrible weapon of mass destruction? Put simply, Snow was saying that a scientist cannot be a scientist and a soldier at the same time, because to use science for war is a refutation of what the scientist is supposed to stand for. Perhaps he had in mind the career of Zuckerman himself. A gentle mathematician who reputedly had a horror of killing even an insect, he spent much of World War II devising complex mathematical formulas for the Royal Air Force that deduced how many German workers' homes needed to be destroyed and how many of the workers and their families had to be killed to wreck German war production. It proved to be a terrible miscalculation: despite hundreds of thousands of German civilian casualties from British bombing, German production actually increased.

This moral debate would arise again, much later in the war and thereafter, but in the meantime it was pushed far into the background as the great fusion of American and British science went forward: the scientists flocked to the colors; the money began to flow into laboratories, universities, and research institutes; and the two great democracies prepared to turn science into an instrument of war, with a speed and power unmatched in history.

Given the tens of thousands of scientists, technicians, and engineers enlisted, the very cream of science and technology underwritten by what appeared to be a bottomless reservoir of money, it is no surprise that the war propelled science faster than even the most radical seers could have predicted. By 1942 it was clear that advances in a host of scientific disciplines were proceeding so rapidly, it was getting hard just to keep track of them all. In a speech to the chemical industry that year, M. A. Stine, scientific research adviser to the chemicals firm E. I. DuPont DeNemours & Company, said, "The war is compressing into the space of months scientific developments which, without the spur of necessity, might have taken a half century to realize."

Even that striking assertion turned out to be something of an understatement. Just to cite three examples, consider how the union of science and war brought into being the airplane, the rocket, and the computer, science's most dazzling contributions to modern civilization—and, in many ways, the most disturbingly ambiguous.

• • •

Mankind's impetus to fly like the birds is as old as mankind itself, but all early attempts to equip men with wings for flight failed. Why that was so didn't become clear until 1680, when the Italian mathematician Giovanni Borelli figured it out: human pectoral muscles were simply inadequate for flight, which birds accomplish with nature's special design that allows them to generate sufficient lift. Humans, Borelli concluded, would need about twenty times their existing pectoral strength to match birds. His solution, which set humans on the wrong flight path for the next two hundred years, was that they must lighten their bodies so that they floated in the air. It was Borelli's conclusion that inspired the Montgolfier brothers in 1782 to invent the lighter-than-air balloon as a means of aiding besieged Spaniards in Gibraltar. War had led to the creation of the balloon, but nobody could figure out how to make it a true weapon of war until 1862, when a young German army officer who had studied aerodynamics got a bright idea.

Captain Ferdinand Zeppelin was serving as a German official observer with the Union army when he watched American balloonist Thaddeus Lowe float his balloons over Confederate lines, and via a telegraph wire stretching back to the ground, report Confederate troop strengths and dispositions. Suppose, Zeppelin wondered, if those balloons somehow could be configured to drop explosives on enemy troops; such balloons would represent the ultimate weapon for any nation. Determined to make Germany that nation, Zeppelin returned home at the end of the Civil War and set to work inventing a true war balloon. Backed by modest funds from a moderately interested German War Ministry, Zeppelin finally came up with a huge, cigar-shaped balloon he called an "airship." It incorporated a number of striking innovations—a lightweight steel frame to hold the gas bags in place, an outer covering of rubberized material so the airship could operate in any kind of weather, use of helium or hydrogen (lighter elements than oxygen, obviating the need for constant heating to expand the gas), and a set of engines for maneuverability and long range.

But the airship remained fundamentally flawed as a military weapon, despite its use to bomb London during World War I. It was aerostatic—its weight was entirely supported by the buoyancy of the helium—and that meant the bombload was severely limited. Moreover, airships were slow and highly vulnerable to antiaircraft fire (a single incendiary bullet, as the British discovered, would instantly turn an airship into a fireball). For a true military weapon, what was needed was something truly aerodynamic, meaning a vehicle that would be supported by the movement of air. Beginning in the late nineteenth century, a number of scientific minds turned to this task, but the solution finally came to two bicycle mechanics named Wilbur and Orville Wright.

Although the Wrights had spent most of their lives in the humdrum world of bicycle gears, they also devoted a great deal of their time to the study of aerodynamics, an effort that led them to two conclusions: One, the airship as a means of manned flight was a scientific dead end; and two, Leonardo da Vinci and Daniel Bernoulli had unwittingly provided the key to manned flight. Da Vinci, the Wrights learned, had researched the flow of water by tossing grain seeds into the currents of a river with a strong current, observing how quickly they moved and in what pattern. His conclusion: The flow of water was more rapid in the narrower section of the river; the river always flowed faster when squeezed through a bottleneck. What did that have to do with flight? Everything, because in 1734, the brilliant Swiss mathematician Daniel Bernoulli discovered that gases and liquids follow the same laws of flow: where the flow is faster, the pressure is lower. When air particles, which tend to keep together, encounter an obstacle, they have to go much faster "around the bend" to keep up with each other.

With this insight, the Wrights built a crude wind tunnel to test shapes for which the flow of air over the obstacle would be faster on the top than underneath, creating "lift." Finally, they found it: a wing with a curved forward edge to direct airflow. On December 17, 1903, a rickety machine they called an "airplane," with curved forward edge wings, lifted off into the wind at Kitty Hawk, North Carolina, and stayed aloft for twelve seconds. At last, man had learned to fly.

From the first moment, the Wright brothers understood the military implications of their amazing invention. They initially tried to sell it to the U.S. War Department, which was distinctly unenthusiastic. Then, in what they claimed was an attempt to keep the airplane an American monopoly, they sold manufacturing rights to European countries while insisting on keeping technical details secret—a fatuous approach, since European science was perfectly capable of figuring out how the plane worked. And once the Europeans understood, they went furiously to work attempting to make the airplane an effective military weapon. In 1911, during Italy's invasion of Libya, the Italians used airplanes for reconnaissance (including the first crude aerial photography) and to drop bombs on Libyan tribesmen.

But it was World War I that provided the major spur to aviation and thoroughly militarized it. The drive to make airplanes more deadly—machine guns, aerial bombs, bigger engines, lightweight airframes, radio communication, electronic navigation—represented a scientific and technical process that leapfrogged what would have been several generations of gradual development and compressed them into a few years. The process illustrated how the pace of technical change was accelerating under the influence of war. By 1916, only thirteen years after the Wright brothers' historic first flight that soared just a few feet off the ground for a few seconds, the Germans were flying the Gotha bomber, a plane with a 77-foot wingspan, powerful twin engines, an operating ceiling of 21,000 feet, a Zeiss vertical telescope bombsight, 660 pounds of bombs, and an operating range of several hundred miles.

No one with even a rudimentary knowledge of aerodynamics doubted that the airplane would someday revolutionize transportation, but few could have guessed how fast that process would occur. The reason for that speed was solely military: the concentration of scientific and technical resources in two world wars to improve what every industrialized nation was convinced represented the new ultimate weapon. Vast treasures were poured into sprawling research and testing facilities, laboratories, and design bureaus for the wide range of scientific and technical developments necessary to perfect the air weapon, a weapon that required only slight modifications to be adapted for civilian use (unlike the tank, which required extensive modifications for use in such machines as bulldozers).

The extensive effort to perfect the air weapon gave birth to a long list of new technologies, chiefly lightweight metal alloys; turbojet engines; pressurized cabins; airborne radar; all-weather electronic navigation systems; autopilots; flight guidance systems; and, in what would prove to be the most revolutionary of all, the solid-state transistor, developed to replace vacuum tubes aboard B-29 bombers that tended to shake loose at high altitudes and malfunction in temperature extremes. (The furious pace of technological development also gave birth to more mundane results, including the ballpoint pen, invented to replace the fountain pens navigators used to mark their maps; the pens squirted ink all over the maps at altitude, moving navigators to demand a new kind of pen that could work at any air pressure.)

There was, of course, one additional scientific breakthrough that would make possible the age of widely accessible air travel and render the world smaller. Like the airplane, the computer was a product of war, but its development followed a more intriguing path.

• • •

In 1940, the brilliant British number theorist G. H. Hardy wrote a slim volume entitled *A Mathematician's Apology*, a classic of mathematics in which he was proud to say of his tiny universe of pure mathematics that "very little of mathematics has any practical utility . . . real mathematics has no effect on war." Hardy went on to talk about pure mathematicians as rarefied human beings involved in an aesthetic endeavor, a pursuit of beauty and order. Presumably he meant to include the more brilliant of the mathematical students he taught at Cambridge, the ones who gathered around his brilliant flame, like adherents of a divine cult.

But Hardy did not know that these brilliant students—including the acknowledged brightest star, Alan Turing—had already been enlisted for war. He did not know because the students had been enlisted two years before in the deepest secrecy for an assignment so sensitive and so vitally important, a mere whisper of it would jeopardize what would become the greatest secret of World War II. Under pain of death, the mathematicians could never breathe a word to anyone about what they were doing, which, they were informed, involved nothing less than winning the war.

It would have been difficult to conceive of any military implication in "pure" mathematics, where mathematicians in this rarefied realm pursued such esoterica as prime numbers and number theory. For many hundreds of years, military science was strictly applied science, and the kind of mathematics important to war involved practical mathematics—such as calculus, critically important in ballistics. But World War I touched off a communications revolution. To move and direct mass armies in a timely way, the telegraph was essential to dispatch orders. Concealing those orders from prying eyes meant they had to be enciphered, setting off a race between codemakers and codebreakers. The heavy traffic necessary to coordinate all those armies and their logistics quickly outran the capacities of code clerks to encipher this huge volume. The next step was inevitable: ciphering machines, electronic devices that could handle all that traffic and simultaneously encipher and transmit it, using interior wiring to generate random ciphers. The most advanced cipher machine, developed in the 1920s, was the German Enigma, which featured a system that could generate up to twenty-six million different possibilities on every keystroke.

That would have appeared to end the work of codebreakers armed with paper and pencil, but two codebreakers—William Friedman, head of America's Army Security Agency, and Marian Rejewski, chief of the Polish army's codebreaking bureau—simultaneously arrived at a crucial insight: The machines were vulnerable to attack if their wiring systems could be divined from analyzing the signals they generated. But only mathematicians could accomplish such a task, which amounted to mathematically figuring out how the machines' electronic signals worked to allow recovery of the cipher "keys" (a sequence of letters or numbers that determined the settings, changed periodically). Friedman and Rejewski recruited mathematicians and discovered that "pure" mathematicians, accustomed to dealing with such intricacies as number sequencing, were perfectly suited to attack complex electronic systems. Friedman's mathematicians would go on to crack the Japanese "Purple" coding machine, giving the United States a priceless advantage in the Pacific war, and Rejewski's mathematicians made significant inroads into the German Enigma.

Following the fall of Poland in 1939, Rejewski gave the fruits of his labors to the British, who immediately realized he had provided at least the beginning of what could be the greatest weapon in the history of war: the ability to read an enemy's enciphered communications. As the Poles had learned, all of Nazi Germany's top-level military communications, including orders from Hitler, were dispatched via Enigma. But it would not be enough simply to crack a message days, weeks, or months after its dispatch; the only valuable codebreaking effort would be able to crack the messages fast enough to use the results immediately. If, for example, Enigma dispatched a message that the Second SS Panzer Corps had been ordered by the Wehrmacht high command to attack in such and such direction on such and such date, it would be vital to crack that message at about the same time the unit received it, so commanders on the ground could organize a counterattack. Scientifically, that amounted to a tall order; some kind of mathematical system would have to be devised to uncover the "keys" for any given message, along with the contents of that message—and it would have to be done in what the technicians called "real time," meaning immediately.

Alan Turing figured out how that could be done. Turing was among a group of Britain's top sixty mathematicians recruited by the government for an all-out attack on the Enigma machine. Like the others who agreed to join, Turing, a British patriot who saw no problem in using his scientific ability to aid his country at the moment of greatest danger, signed a draconian security document that forbade him to discuss his work in his lifetime with any outsider, even his own family. In great secrecy, Turing and the others were taken to Bletchley Park, a country estate just outside London turned into a sprawling codebreaking operation that eventually would contain more than twenty thousand people working around the clock in shifts.

As a class, mathematicians—especially pure mathematicians—tend to be an odd lot, and the oddest of the group at Bletchley Park was Turing. A Cambridge undergraduate whose mathematical brilliance made him a protégé of Hardy, Turing wore a gas mask each spring to protect his delicate respiratory system from hay fever, chained his tea mug to a radiator to prevent what he was certain was a massive conspiracy to steal his teacups, and often was so absentminded, he appeared for work dressed in pajama tops. But his intelligence was in perfect order, focused like a laser beam on the pure mathematician's obsession with absolutely logical deduction.

Among other things, for some years Turing had been thinking

about something he modestly called a "Turing machine," basically an electronic calculator that could instantly perform virtually any mathematical calculation on the basis of a stored program of all possible mathematical results. Such a machine, in common use today, would have to await the invention of small electronic circuitry to make it practical, but the concept provided him with a vital insight into the Enigma problem. As Turing reasoned, although Enigma had several rotors whose settings, changed periodically, could generate many millions of possible letter combinations with each stroke of the machine's keyboard, the total number was still finite. What if an electronic machine could be built that would store all the possible combinations? An Enigma message would be fed into the machine, which would then be instructed to search its memory bank to recognize which of the many possible settings had been used for enciphering. With that crucial foothold, the message could be deciphered.

Turing designed a large machine 17 feet long and 8 feet high, powered by some 2,500 vacuum tubes, that could read 2,500 characters a second. He called it a *bombe* (in honor of the Poles, who had tried, unsuccessfully, a similar approach and named it after a popular brand of Polish ice cream). By the end of 1940 it began to break into a wide range of Enigma messages, and by war's end, the overall deciphering operation, code-named ULTRA, was reading messages even before the intended recipients received them.

While Turing was building his *bombe,* the other components that collectively would create the postwar computer revolution were coming together under the impetus of war. University of Pennsylvania mathematicians developed the Electronic Numerical Integrator and Computer (ENIAC), a monster machine weighing 30 tons that consumed 200 kilowatts of power every time it was turned on, enough to dim the lights of the entire city of Philadelphia. It was designed to help the army with the daunting task of making all the voluminous mathematical calculations needed for the firing of anti-aircraft guns (it cut the average calculation time from twenty hours to thirty seconds). At Harvard, two mathematicians working for the navy built the Mark I, a 50-foot-long machine with 530 miles of wire that instantly calculated naval gunfire ballistics with such variables as speed of the ship and sea conditions, fed into the machine with a punched-tape programming system, the prototype of computer software. These wartime machines would have to await several scientific breakthroughs before they could achieve real practical use and their full potential—most importantly the improved transistor in 1948 and, ten years later, the integrated circuit, a chemical miracle made from quartz rock that combined three electronic components onto a small silicon disc. (These two essential developments grew out of military programs, chiefly the need for high-scale computing to calculate such mathematically daunting tasks as the design of missiles and nuclear weapons.)

The highly successful utilization of mathematicians by Great Britain and the United States for war was not duplicated in Germany, which had plenty of its own brilliant mathematicians. But the crucial difference was that Allied science was organized, and Germany's was not. The British and the Americans had a much firmer grasp of how to use science and technology for winning wars. That advantage centered not on the science itself, but on how it was organized for maximum effect. For all the talk about German efficiency and organization, the truth is that Nazi Germany was a hopelessly disorganized mess, especially in science and technology.

Consider the matter of computers. In 1941, a German mathematician, Konrad Zuse, built an electromechanical relay computer that used an IBM punched-card system. Zuse developed the machine on behalf of the German aeronautical industry, which wanted to save the large amount of time needed (using conventional mechanical desk calculators) for such mathematically laborious calculations as drag vortexes for wing design. This early computer saved airplane designers a lot of time, but that's as far as it went. Under Germany's strict hierarchical system, nobody in the aircraft industry thought to tell the government that a marvelous machine now existed, one that might be used for other war-related tasks. Among such tasks, obviously, would be codebreaking, but the Forschumgammt, Germany's largest codebreaking organization, was run by the Luftwaffe, and its mathematicians, sweating over Allied codes with paper and pencil, were not informed of Zuse's computer. Nor were codebreaking organizations run by the SS and other German military organizations, each of which operated in its own separate realm, oblivious to whatever scientific or technical developments were occurring elsewhere.

The German research and development effort was badly fragmented, isolated from the world of military operations and planning. German scientists, scattered all over the military and industrial establishments, worked uncoordinated, concerned only with the narrow interests of whatever military arm or industry was paying their salaries. Too many of them were busy trying to invent the weapons of World War III—such as a group of scientists working for the Luftwaffe who wasted a lot of time, money, and effort to come up with a long-range strategic bomber capable of dropping bombs on New York City. A prototype was built and actually flew within a few dozen miles of the city in a secret test flight, but by that time it was late 1944, and such a weapon was pointless. Even assuming that more than one model of the plane could be built by Germany's beleaguered aircraft industry and drop a few bombs on New York City, so what? It would not have put even the slightest dent in America's industrial capacity, which was in the process of burying the Nazi war machine under mountains of weapons. But the project had gone forward because of the prime determining factor in German science, Adolf Hitler.

All matters large and small in Germany ultimately devolved on Hitler, who reserved the right to decide which areas of scientific research would be pursued and which weapons systems would be developed. Hitler, however, was a scientific and technical illiterate whose insistence on judging all scientific matters guaranteed that the German scientific effort would operate in fits and starts, subject to his whims. Even when German science came up with a striking advance, there was at least a good chance that Hitler would foul it up. Among the more prominent examples was a great triumph of German science, the development of the world's first operational jet-powered fighter, in 1943. Hitler had approved its development, but when finally presented with the finished result, decreed that the plane must be produced as a fighter-bomber—which, as appalled German aeronautical scientists realized, meant that the plane's huge advantage in speed would be thrown away: a jet plane that had to carry bombs and fly slowly for bombing runs defeated the whole idea of the jet plane in the first place, which was to shoot down all those Allied bombers.

Nevertheless, German science was still able to come up with a number of amazing advances, among them its most glittering achievement during World War II: rockets (although this weapon, too, would be frittered away by disorganization, indecision, and the time lost waiting for Hitler to make up his mind).

The German achievement was actually born in the United States in 1919, when a Clark University physicist named Robert Goddard solved a knotty problem that had puzzled science for nearly seven hundred years. The Chinese invented rockets early in the thirteenth century, and although there were plenty of military thinkers and technicians who dreamed of giant rockets that could be launched to hit targets hundreds of miles away, no one could figure out a way of overcoming their inherent limitation—gunpowder, the propulsion system. The problem was that no matter how much gunpowder was used, it could not generate sufficient power to move a heavy rocket any significant distance. (Even William Congreve, a British army officer who developed bombardment rockets in the eighteenth century and was a fanatical proponent of their use to replace artillery, conceded that his best rockets were inaccurate and could not carry heavy payloads of explosives—as demonstrated in 1814, when he bombarded Fort McHenry with hundreds of rockets for an entire night, only to discover the next morning that the American flag was still defiantly flying. This failure is immortalized in the "rocket's red glare" line of "The Star-Spangled Banner.")

To make the rocket propulsion problem even more complex, there was an additional issue that arose in any attempt to launch a rocket beyond the earth's atmosphere: Gunpowder would have no oxygen to burn. Goddard believed this problem could be overcome and a way could be found to launch a heavy rocket into space. He began by studying why even a small rocket powered by black powder didn't go very far, finally concluding, after mathematical analysis, that any form of explosive powder, no matter how powerful, could never produce sufficient "boost" (the number of pounds of thrust produced by each pound of propellant burned per second). Goddard went on to a scientific insight that would finally land man on the moon. The solution to the propulsion problem, he concluded, was that the rocket must carry its own oxygen supply, a liquid version combined with a fuel that has a very high and powerful burn rate, such as hydrogen. Goddard built a few small-size rockets that proved his concept, a result that led him to predict 10-ton rockets with multiple stages (another Goddard innovation) that would escape the atmosphere and land on the moon.

Goddard's interest centered strictly on space rockets, an idea that excited not even a flicker of interest in America, save a *New York Times* editorial (apparently written by someone who had flunked eighth-grade science) criticizing Goddard for overlooking a basic principle of physics, namely that a rocket could not operate in space because of a lack of oxygen. The reaction was quite different among a small group of German rocket enthusiasts who excitedly read his brilliant scientific paper "A Method for Reaching Extreme Altitudes" and adopted his technical ideas for their own rocket experiments. But in 1935, the rebuilding German army, desperate for new long-range artillery, enlisted the enthusiasts—including a graduate physics student named Wernher von Braun—to develop long-range ballistic rockets capable of carrying large explosive warheads.

Eventually von Braun's group developed the V-2 rocket, precursor to the Space Age, a brilliant design that produced 28 tons of thrust from a fuel of liquid oxygen and alcohol, a set of gyroscopes, and flight guidance fins, all to launch a 400-pound warhead of high explosives on a target hundreds of miles away. A very impressive scientific and technical achievement, but like so many other fruits of German science, it was never able to achieve the status of decisive weapon. In the disorganization of German wartime science, there was no central directing authority, as in the United States and Great Britain, to evaluate scientific advances and then prioritize resources to bring them to fruition. Instead, the developers of the German ballistic missiles had to wait as a vacillating Hitler finally made up his mind whether he would order enough resources to develop long-range rocket weapons. Precious months passed, and by the time Hitler made his decision it was 1944, when the war was already lost for Germany. The V-2 rockets that landed on London were terrifying, but aside from blowing up some buildings and killing several hundred civilians, they achieved absolutely nothing militarily.

Other German rocket development programs produced equally advanced weapons, notably the Fritz-X, the world's first cruise missile. The missile was carried by a bomber to within several miles of its target and then launched, guided to its target by "riding" radio waves emitted by the bomber crew. Unleashed in 1943 against U.S. and British ships in the Mediterranean (one Fritz-X blew the gun turret off an American cruiser), the missile was initially successful, but checkmated within just a few months by Allied countermeasures, which involved jamming the radio signals that guided the missile.

The Allies' ability to come up with a counterpunch to the Fritz-X so quickly underscored a major scientific advantage they had over the Germans: a centralized scientific research and development apparatus that could react quickly to any new threat, redirecting resources to come up with a counter. Repeatedly, a German scientific advance that had taken years to develop would be stymied in months—"chaff" (millions of thin strips of tinfoil) to confuse advanced German radars; depth charges hooked up to sonar that exploded only when a U-boat was found, defeating new deep-diving German submarines; a shaped charge fired by a rocket launcher called a "bazooka" to penetrate new, thick-armored German tanks; an electronic "field" around ships to balk German antiship mines set to explode when they detected the vibrations of a passing vessel; a long-range fighter called the P-51 Mustang, conceived and developed in only eight months, to protect bomber formations over Germany from enemy fighters.

There was another Allied scientific advantage: organized scientific intelligence. The Germans didn't have it, and that omission would prove very damaging to their war effort. Assuming that their science was superior and that their scientists would always know of the latest scientific advances, the Germans never bothered to create a mechanism to collect scientific intelligence. That oversight would cost them dearly in the Battle of Britain, when advanced British radar was able to detect incoming German bomber and fighter formations up to several hundred miles away, at the same time determining their speed, heading, and numbers. The Germans could not say they hadn't been warned: In 1938, the British began building a series of large, mysterious towers along their eastern coast. The Germans assumed the towers were connected to radar in some way, and sent the airship *Graf Zeppelin II*, ostensibly on a pleasure trip, to "accidentally" stray off course and record signals the towers were emitting. German physicists who later listened to what the airship recorded pronounced the British system "useless," since it broadcast at higher frequency ranges; the German scientists working on radar believed radar was feasible only in lower frequency ranges. The Germans didn't know that the British had developed radar that could operate at higher frequences (and thus longer distances) for the simple reason that they had no scientific intelligence unit that kept track of scientific developments outside Germany—especially those with military applications.

But the British and the Americans did. Even before he became prime minister, Churchill was running an ad hoc intelligence unit of industrial experts and scientists he knew to report to him anything of scientific interest they may have encountered. As a result, when Churchill took office, he had a fairly comprehensive overview of the state of German science. He ordered British intelligence to retain a science adviser who could evaluate scientific intelligence. The man chosen, the physicist Reginald V. Jones, turned out to be perfect for the job. Among other feats, he discerned the existence of the German V-2 program after noticing an unusual German interest in liquid oxygen, and discovered that German bombers were using a radio direction beam to find British targets (a beam Jones caused to have misdirected by an even stronger British beam, resulting in the bombers missing their targets). The Americans had a much more elaborate scientific intelligence operation that formed the core of the Research and Analysis Branch of the Organization for Strategic Services (OSS). The unit, run by prominent academics, reached deep into the faculties of hundreds of colleges and universities, tapping into a wide range of scientific and technical expertise to draw useful information from even the most mundane clues (such as the metals expert who, just by examining a German ball bearing, could determine where it was made, the technical level of the machine tools its manufacturing process used, and the quality of the metallurgical process that produced its steel).

The creation of organized scientific intelligence underscored just how pervasively science had penetrated all aspects of war. There was virtually no field of science that could not be used in the war effort and not a single scientist who could not contribute in some way—mathematicians to build computers and solve gunnery problems; chemists to develop new explosives; biochemists to develop new medicines for troops vulnerable to such tropical diseases as malaria when operating in jungle areas; physicists to determine the most efficient bomb patterns; psychologists to develop better troop training methods and select officers; oceanographers to determine sites for amphibious landings; and meteorologists to make accurate weather predictions for military operations. In the process, the stimulation of war set off the greatest growth in science and technology in the history of the world. It would require a separate volume merely to summarize this vast transformation, which ranged from the momentous (nuclear weapons) to the relatively mundane (frozen orange juice concentrate, developed to prevent scurvy among front-line troops and naval personnel). Almost every aspect of modern industrialized society owes its origins to science's role in two world wars—aerosol cans, developed so troops in tropical areas could spray antimosquito insecticide in tight spaces; synthetic rubber, developed to replace natural rubber supplies cut off by the Japanese seizure of Malaya; microwave ovens, accidentally discovered when a scientist walked past a cavity magnetron for radar and noticed that a chocolate bar in his pocket melted; nylon, developed to replace silk in parachutes; freeze-dried foods, developed to provide combat troops with lightweight, nonspoilable food rations; telephone technology, developed to make communications between artillery observers and artillery batteries more efficient; and high-performance truck diesel engines, developed to power multiton tanks.

Among the more striking aspects of this catalog of science and technology devoted exclusively to the art of killing other human beings is the near-total lack of uneasiness among the thousands of scientists involved. There were no petitions to the government from scientists protesting the use of science for destruction and no organizations of concerned scientists advocating science's withdrawal from participating in the conduct of a war that in the end would be responsible for the deaths of nearly thirty million human beings. That had much to do with both the nature of the war and the nature of the enemy. There were very few people who did not understand that Nazism was a great evil, an evil so malignant, so great a threat to civilization, that there was little that could not be justified in the name of eradicating it. So was Japanese militarism; it would have been difficult to find more than a handful of sympathizers for the rapists of Nanking. There were no recorded crises of conscience in the mind of the British zoologist who killed hundreds of monkeys in tests he conducted to determine the effects of high-explosive bombs on flesh and bone. His work, honored by a grateful British government, was eminently justifiable: with his data, bombs could be constructed to more efficiently kill hated Nazis, the evil monsters of concentration camps, slave labor, and slaughter of innocent civilians. And there were no known crises of conscience among the Harvard chemists who developed napalm, a terrifying weapon designed to incinerate enemy troops in deep fortifications or caves. Certainly it must have occurred to them that the weapon was indiscriminate; it literally incinerated every living thing within an elliptical area of 30 yards by 90 yards, the mixture of gasoline and rubber sticking to flesh like glue and burning holes in human bodies. But the justification was obvious: to save Allied lives that otherwise would have been lost in assaults against German troops protected by several feet of concrete, and Japanese soldiers in thick bunkers and caves.

But a crisis of scientific conscience did take place, and the moment of its greatest intensity can be set with precision. On August 6, 1945, at exactly 8:16 A.M. Eastern Standard Time, a single American B-29 bomber arrived over the Japanese city of Hiroshima. The citizens of the city, assuming the lone plane was either on a reconnaissance or weather mission, did not run to air raid shelters.

The bomb bay door on the plane opened, and a 10-foot-long bomb 2.5 feet in diameter began to fall to earth. Inside the bomb, 50 pounds of U-235, encased within 10,000 pounds of cordite and tampered steel casing, waited for an electrical signal. At an altitude of 1,900 feet forty-three seconds later, an electrical impulse lighted several cordite sacks, setting off an explosion that pushed a segment of the uranium down a 4-foot-long barrel. Neutrons were shoved into the nuclei, which began to jostle and wobble, beginning a chain reaction that reached critical mass, with temperatures of 5,000 degrees Centigrade. A millionth of a second later, the bomb buckled and bulged, followed by a blast of unprecedented power, accompanied by an intense white light. On the ground, some 127,000 human beings died and 60 percent of the ancient city crumbled into broken stone and dust.

The next day, a shocked *L'Osservatore Romano*, the official Vatican newspaper, wondered how the consciences of the scientists involved in developing the weapon could have allowed them to construct something so evil. Why, the newspaper wondered, hadn't the scientists followed the example of Leonardo da Vinci and refused to share their knowledge of "destroying engines"?

That would turn out to be a very good question.

8

A Thousand Suns

*The physicists have known sin, and this is a
knowledge which they cannot lose.*

—J. ROBERT OPPENHEIMER

By Christmas morning of 1938, the snowstorm had ended, and the woods and the fields just outside Stockholm were clothed in a bright white blanket, the chill hush broken by the happy laughter of children testing out their Christmas presents of sleds and skis on the low hills. Several adults were out for a walk, relishing a holiday morning of bright sunlight glistening on the snow, a rare sight in the city's normally gloomy winters.

Two of those adults seemed not to notice the bright sunlight, or the snow, or the children playing. Heads down, they grimly trudged through the snow as they talked in rapid-fire German. Even if any of the passersby understood German, what this man and woman were discussing with such obvious emotion would have been unintelligible, a strange dialect of neutrons, nuclei, protons, particles, elements, and isotopes. Clearly, they were both unhappy and agitated, as if they had just learned of a terrible danger looming over their lives.

In fact, they had. Otto Frisch, a nuclear physicist, had rushed to Stockholm to meet his aunt, Lise Meitner, also a nuclear physicist, and discuss what they both agreed was an event so fraught with danger and ominous implications, the likelihood was that this Christmas would be the last they would ever share. The source of their concern was an event that only a small number of people in the scientific community were yet aware of: A few weeks before, two German chemists, Otto Hahn and Fritz Strassman, had conducted an experiment in which they bombarded a shielded piece of uranium with neutrons. This procedure resulted in a puzzling result: Each nucleus of the uranium atoms broke into two smaller pieces, roughly equal, that flew apart with a tremendous amount of energy and produced more neutrons. The atoms that resulted weighed slightly less than the ones that went into the process. Chemically, that wasn't supposed to happen: Where was the missing mass? As chemists, Hahn and Strassman assumed they must have made some sort of mistake in interpreting the results.

Hahn contacted Meitner, a former colleague, to convey news of the puzzling experiment, and although he couldn't figure out what had happened, Meitner did. She said nothing about her conclusion to Hahn but subsequently contacted her nephew, who was working in Copenhagen with the famed nuclear physicist Niels Bohr. Disturbed, Frisch rushed to Stockholm to discuss what he and his aunt realized was a momentous event, both scientifically and politically. On Christmas morning, following a night of intense discussion, they decided to work off their restless uneasiness by taking a walk.

As they walked along in the snow, Frisch and Meitner agreed that Hahn and Strassman hadn't made a mistake. By the time they decided to rest and sit on a tree stump, both realized that in fact the two German chemists had stumbled onto the force that could destroy the world. The result did not come as a total surprise to Frisch, since Bohr two years before had speculated that the nucleus of the uranium atom appeared to be quite weak, and might deform or rupture from the impact of even the tiniest particle, such as a neutron. Further, he speculated that if the nucleus was bombarded with enough neutrons, it might begin to jiggle more and more wildly, until finally the forces that held the nucleus together gave way. Then the forces inside would make the fragments of the nucleus fly apart, releasing tremendous energy. Ignoring the biting cold, Meitner and Frisch pulled out pencils and papers from their pockets and began scribbling equations. When they finished, they looked at each other in somber silence for a moment. There was no doubt: The two Germans had split the atom and in the process converted mass into energy, fulfilling Einstein's famous mass-energy equation. The ages-old scientific principle that matter cannot be created nor destroyed had been proven wrong by two chemists who didn't realize what they had done.

But this was not simply a matter of nuclear physics. The rough mathematics that Meitner and Frisch had scribbled while sitting on the cold stump were irrefutable—splitting atoms on a much larger scale than the Hahn-Strassman experiment would unleash a burst of energy beyond human comprehension, a cataclysmic explosion to dwarf anything in mankind's experience. And if that power could be converted into a weapon, the result would be the plus ultra of weapons.

The moroseness of Meitner and Frisch as they contemplated the implications of their scribbled mathematics was related to the political context. Hahn and Strassman were loyal Germans who had gone along with the Nazification of German science. Their *Führer* had ripped up the Versailles Treaty, which limited German rearmament; had forced a union with Austria; and had extracted a chunk of Czechoslovakia. Clearly he had further aggressive intentions, judging by the massive German military buildup. German science, obviously, would be just another weapon in his arsenal, a science that at some point would realize that Hahn and Strassman had not made a mistake and then undoubtedly seek to convert what they had discovered into a frightening weapon. Meitner, who had been driven out of Germany because she was Jewish, was fully aware of the dark forces that Nazi Germany represented; she had lived through the nightmare of Nazi storm troopers strutting around the streets, the heavy hand of repression, the militarization of every aspect of German society. She did not need to be told the implications of an ultimate weapon in the hands of a man like Adolf Hitler.

Now what to do? They decided that Bohr, among the world's most preeminent nuclear physicists, should be told at once. Frisch rushed back to Copenhagen and managed to catch Bohr just as he was about to board a ship to travel to the United States for a theoretical physics conference in Washington, D.C. Slightly out of breath from his dash to the ship, Frisch broke the news to Bohr, who listened silently with furrowed brow, the certain sign that his magnificent mind was already absorbing the implications of what he was being told. Searching for a word to describe just what Hahn and Strassman's experiment achieved, Frisch finally remembered a term from a biology class of his school days, used to describe cellular division: fission.

"Fission," Bohr repeated, nodding, then boarded the ship. Some days later he electrified the physics conference with an analysis of the Hahn-Strassman experiment and its implications. There were excited murmurs all over the room as Bohr described the fission process and what he called "chain reaction," the phenomenon that occurs when the uranium nucleus splits apart into two large fragments, plus two or three neutrons—which, if they strike another nucleus and induce still more fission, results in three more neutrons that create a total of nine neutrons, and then on and on until all that fission sets off a huge release of energy in millionths of a second.

There was no necessity for Bohr to spell out the immense military implications, because every physicist in the room instantly understood that if the chain reaction could be controlled in some way, mankind would achieve the terrible Promethean fire of Greek myth. And their conclusion was unanimous: Such a weapon in the hands of Adolf Hitler threatened the very existence of the planet.

Two of the conference attendees, both European émigrés who like Meitner knew firsthand the danger of Nazism, immediately set to work. Enrico Fermi was determined to figure out a way to control the nuclear chain reaction, while Leo Szilard had a more ambitious plan in mind: Somehow, in some way, the United States must seek to convert the fission process into a weapon before Nazi Germany did.

A few months later, Hahn and Strassman published a paper about their experiment in the prestigious German science journal *Die Naturwissenschaft*. They did not know that Germany's leading scientific publisher had obtained a copy of the manuscript before publication and slipped it to a British intelligence agent. In London, the copy was read by an alarmed group of physicists the British government had recruited for scientific intelligence work; they immediately recommended that Britain begin working on a weapon that would harness the power of fission. In Moscow, the Soviet Union's leading nuclear physicist, Igor Kurchatov, read the Hahn-Strassman article with growing excitement. A few hours later, he personally recommended to Joseph Stalin that work begin at once on a Soviet fission bomb.

By then, Hahn and Strassman had realized the implications of their experiment, as did German nuclear physicists, who became concerned when the Hahn-Strassman article attracted virtually no reaction outside Germany. The lack of reaction was telling: Undoubtedly, at least several nations were working on a fission bomb.

It was time for Germany to make its own bomb. Several months later, in April 1940, Frisch fled from Copenhagen ahead of what he assumed was an imminent German invasion and went to Britain. Recruited into the British scientific war effort, he joined a small group of nuclear physicists; his first action was to prepare an outline of how a practical fission bomb could be built—if, he noted, the required amount of resources, time, and scientific talent were applied to the task.

And with that, the stage was set for what would become the greatest scientific drama in history, a penultimate union of science and war that for the first time gave mankind the power to destroy itself. It would cause science's gravest crisis of conscience, for never again could science claim that it is strictly an objective reality independent of laws, customs, authorities, and loyalties, a pillar of objective thought divorced from the axioms of society. The atomic bomb was conceived and built strictly by science; only a small circle of nuclear physicists understood that an infinitesimally small bit of matter contained enough energy to blow the planet apart. And it was the scientists who decided to join that knowledge to the military institutions of the state; no one forced them to do that.

• • •

By now, the story of how science discovered the structure of the atom is well known—the discovery of X rays by Wilhelm Roentgen in 1895, the discovery by Ernest Rutherford in 1911 that the atom has a nucleus, and by 1930 the various discoveries that showed the atom as a miniature universe, a dense nucleus of protons and neutrons surrounded by a diffuse cloud of electrons (protons carry a positive electrical charge, electrons have a negative charge, and neutrons have no charge). How all this was discovered represented absolutely pure science, since there was no discernible practical use for it. A small fraternity of nuclear physicists spent their time trying to understand how something they could not actually see worked, an invisible world where the nucleus of an atom was ten quadrillionths of a centimeter across. They regarded themselves almost as a priesthood, the last true scientists devoted to understanding the riddle of the universe, the very purpose that had lured them into science in the first place. They belonged to an international fraternity that transcended national boundaries, openly sharing their research and discoveries. They would later remember those years before Hitler as a glittering age, a time when they were free of the political concerns and ethical dilemmas that confronted scientists in other disciplines. Their golden knight, Albert Einstein, proclaimed, "It is a many-mansioned building, this temple of science." And science, he added, was a "flight from . . . everyday life with its painful rawness and desolate emptiness. . . ."

In a modern context that sounds somewhat naive, and to a certain extent it was; Einstein considered himself a citizen of the world and was largely detached from the politics and the society of his native Germany. But he was also reflecting the mood of the little universe that gave meaning to his life, a universe bathed in the light of reason. It operated in a collegial atmosphere free of international politics; German, American, Italian, Russian, and Japanese scientists met on a common ground where the only loyalty that counted was the loyalty to scientific truth. The German university town of Göttingen was the Camelot of this world, the intellectual center of nuclear physics where physicists from around the world came to study and debate. The citizens of Göttingen had no idea what these people were talking about, but nevertheless treated them with the German deference toward people of great learning, and tolerated some of their idiosyncrasies—the Russian physicist who fell into the gutter and refused to let anyone help him up because it might break the train of thought that had caused him to fall, the German physicists who rushed into a restaurant, demanding to see the tablecloth on which they had dined the night before (and on which they had scribbled equations); the absentminded American physicist who drove his car up the steps of a building while lost in thought. In retrospect, it was a bittersweet memory for those involved, for it was an age of innocence, almost certainly science's last such age. And that age ended, ironically enough, in the place where it was born.

The Germany that evolved from the wreckage of Napoleon's defeat of Prussia was a nation with an ethos of duty, rectitude, and obedience, the essential traits for a society determined never again to be a second-class power. Germany's rapid development of science and technology, along with their first cousin, industrialization, and a near-cult of education created a major world power whose obsession with becoming the dominant land power in Europe provided cohesion for a divided people—a rigid, efficient, hardworking people bent on achievement, to establish Germany's "place in the sun." They had certainly achieved it in science: By 1933, German scientists had won 30 percent of the Nobel Prizes for science, a statistic largely reflective of Germany's first-class educational system, focused strongly on science.

But for all the faith in science, it could not save Germany from defeat in World War I. The Germans learned nothing from the experience, for the defeated armies no sooner had trudged back home that they were busy rebuilding a science and technology they were convinced would lift them from the ashes of defeat. All over Germany, there were laboratories and testing centers involved in secret work to help the German military evade the restrictions of the Versailles Treaty.

It was a development that appalled Einstein, Germany's most famous scientist. His worldwide fame began in 1919, when a British scientific expedition observing a solar eclipse confirmed Einstein's predictions about the behavior of light in his 1905 paper that spelled out the General Theory of Relativity. The worldwide fame was, to a large extent, a product of its time: New heroes were needed to replace the old heroes buried in the rubble of World War I, new heroes untainted by war. And in science, there were very few who had not been tainted by their involvement in war. For that reason, Einstein, a former enemy, could be welcomed with great fanfare in Great Britain, but it was a country where Fritz Haber, the German scientist whose poison gas had killed and maimed so many British soldiers, dared not even show his face.

A dedicated pacifist, Einstein had criticized what he called Germany's "greed" in World War I, and now he saw all the signs of a resurgence, abetted by the German scientific establishment. Worse, there was something else creeping into German science, an odious cancer that had taken root in the unrest and resentment of postwar Germany. Einstein had seen the approaching danger and tried to warn his friend Max Planck, the brilliant physicist whose quantum theory was then revolutionizing physics. But Planck, like many of his fellow scientists, saw no need for alarm. "A passing phase," he said to Einstein of the growing power of Hitler and the National Socialists. Planck would learn the hard way that Einstein was right: In 1933, as Hitler came to power, he personally went to see *der Führer* to plead for the reinstatement of Fritz Haber as head of one of

Germany's most prestigious scientific research institutes (Haber, a Jew, had been forced out by the Nazis). To Planck's shock, Hitler flew into a titanic rage, screaming at Planck that no Jew would ever head a scientific institute in Nazi Germany again, and what's more, wondered why Planck, an "Aryan," would intervene on behalf of "a Jew swine."

With that, a dismayed Planck withdrew from science, deciding he would never let his scientific talent be used for a regime that called a Nobel Prize–winning scientist "a Jew swine." (Unable to bear leaving his beloved Germany, he retired in isolation to his home in Göttingen, there to reminisce about the golden days of innocent science in the city. He died in 1947—of a broken heart, his friends said, having undergone the tragedy of Germany's destruction; his own home and library reduced to ashes by Allied bombing; and, most tragic of all, the execution of his son for his involvement in a plot to assassinate Hitler.)

Einstein fled Germany in 1932, just as the new director of the Institute of Physics in Dresden announced that modern physics amounted to "instruments of Jewry." At the same time, a hundred German professors who had joined the Nazi Party signed a statement that denounced the Theory of Relativity as "unscientific" and flatly wrong. "If I were wrong, one would have been enough," Einstein remarked dryly, from his new home in exile in the United States. It would mark the last time that Einstein would be so flippant, because a dark shadow now fell over German science. Two physicists, Philip Lenard and Johannes Stark, suddenly vaulted to leadership of all scientific matters in Germany solely because they were ardent Nazis. With Hitler's blessing, they weeded out all Jewish scientists and any others believed to be insufficiently sympathetic to the new regime. From now on, Lenard and Stark decreed, there would be only "Nazi science" in Germany. The world would come to know what "Nazi science" meant—anthropologists who worked for the SS to propound crackpot theories of racial superiority; medical scientists who conducted grisly experiments on concentration camp inmates; chemists who developed Zyklon-B poison gas to murder millions of Jews; biologists who propounded a theory of "human natural selection," under which thousands of the mentally retarded were murdered because they were "unfit" to live; geneticists who worked to prove that all Jews and "non-Aryan" races were inferior, thus deserving of extermination or enslavement.

This was the environment in which German science was enlisted to develop the weapon of weapons, the weapon that would guarantee Germany's triumph over all its enemies. That effort would fail, partially because of the state of German science and partially because of the scientists involved in the project—especially the man who headed it. What he did or didn't do remains a matter of fierce controversy to this day.

• • •

Unlike Planck, Werner Heisenberg saw no moral problem in serving an odious regime. Privately, he liked to describe himself as an anti-Nazi, but saw his duty to "save" German science for rebirth after the war certain to come, when the Nazis would be defeated and the "good Germans" would retake control of Germany. For that reason, in 1939, while on a trip to the United States, he turned down offers of asylum from American colleagues. Instead, he returned to Germany and that September, as World War II broke out, he walked into the German Army Weapons Bureau and volunteered his services to head a team of nuclear physicists who would build an atomic bomb. Three months later, he presented a scientific paper on how such a bomb could be built. Army officials told him to go ahead, and Heisenberg, after rounding up a team of a dozen German nuclear physicists, moved into an old government building in Berlin to begin work (the building was called "the virus house" because it had once housed a biological laboratory and because the name served as cover for the deeply secret scientific research that now went on there).

This course of events makes no sense. If Heisenberg was, as he claimed, an anti-Nazi, then what was he doing trying to build an atomic bomb for a regime he professed to despise? No one forced him to work on an atomic bomb; he volunteered to do so. Like Planck, he could have simply opted out of doing anything scientific to help the Nazis. After the war, Heisenberg hinted that it was all part of a deeper game: Aware that the Nazis wanted an atomic bomb, he decided to head the project for the sole reason that he meant to ensure that it never came to pass. The reason why German science failed to build a bomb, he suggested, was that he deliberately stalled and prevaricated the work until the end of the war, to deny Hitler the bomb.

Heisenberg might have gotten away with this version of events had it not been for British intelligence. As Germany collapsed, Heisenberg and his team were captured by the British, taken to Britain, and interned in a country estate called Farm Hall. There Heisenberg and fellow members of what they called "the uranium club" spent their time in debriefings with American and British scientists attached to intelligence; otherwise, they sat around discussing the course of the war and sundry other topics. It was these private conversations that most interested the British, since they assumed (rightly, as it turned out) that the Germans would tell them what they wanted to hear in the debriefings but talk more honestly among themselves. Consequently, the British bugged the Germans' living quarters and recorded the results. The transcripts were not declassified until nearly fifty years later, and they are quite revealing. Above all, they show that the German atomic bomb project did not fail, as Heisenberg suggested, because he had deliberately stalled it, but because German scientists (especially Heisenberg) made just about every scientific mistake in the book. Indeed, their failure was so colossal, so out of variance with the presumed genius of German science, that it has stirred speculation for years that so gross a failure could only have stemmed from deliberate attempts to sabotage the project.

Certainly, Heisenberg had the scientific qualifications to learn how to build a nuclear bomb. Before Hitler came to power, Heisenberg was one of the golden boys of German science, a brilliant physicist who had won the Nobel Prize in 1927 for a singular contribution to nuclear physics, the "uncertainty principle." Essentially, the principle has to do with precisely measuring the most essential properties of subatomic behavior, which, Heisenberg demonstrated, can be done only by observing the position of an electron—and that cannot be accomplished without bouncing something off it, especially light. But in an insight that Heisenberg said occurred to him while he was hiking in the mountains, the mere act of observing makes the behavior of the electron uncertain. Radiation of high energy will provide precise data about where the electron is, but that will destroy the evidence of its initial velocity. The very act of ob-

serving the electron's position will make it behave more like a particle, while measuring its energy will make it behave more like a
wave. Therefore, atomic behavior is indeterminate and cannot be
predicted.

This insight, essential to the study of subatomic particles, vaulted
Heisenberg to the front rank of the world's nuclear physicists, his
work praised by such giants as Einstein, Planck, and Bohr. Heisenberg was one of the leading lights in the golden days at Göttingen,
and in that collegial atmosphere he often could be found strolling in
the woods with other physicists, debating such matters as whether
an electron is a particle or a wave. At such moments Heisenberg
was a scientist first and a German second. But, in a transformation
Heisenberg never really understood, as of 1933, when the Nazis
came to power, he became a German first and a scientist second. To
the Nazis, Heisenberg was a German who happened to be scientist.
And there was now no middle ground: Either he was loyal to the
regime and served it, or he would not serve it and therefore became
an enemy. To a large extent, Heisenberg's tragedy is that he tried to
have it both ways, serving the regime of a man he privately called a
"thug" while claiming that his real and nobler purpose was to preserve science after the evil regime's fall. Heisenberg was not a Nazi,
but working on an ultimate weapon for a regime he professed to regard as a danger to the world made him complicit in its evils—especially since he was aware of those evils.

Like too many of his fellow Germans, Heisenberg had a colossal
moral blindness. He resented sneers among Nazified scientists that
he was a "white Jew" because of his pre-Hitler associations with
such undesirables as Albert Einstein, and his volunteering to work
on an atomic bomb may have been an attempt to prove, somehow,
that he was authentically German in the Hitler era's sense of the
term. As the German atomic bomb project proceeded, he was perfectly aware that brigades of slave labor were being used to process
uranium oxide, a highly dangerous procedure; most of them died.
And when his wife wrote a memoir after her husband's death in
1980, she went to great lengths to justify his actions, claiming he
hated Hitler and that they lived in a form of "inner exile." But there
is not a single mention of the Holocaust in a book that purports to
be an account of Heisenberg's revulsion of Hitler and the Nazis.

In terms of the German atomic bomb project, for all his hints that he secretly sabotaged it, the fact is that the project failed not because Heisenberg ensured it would never come to pass, but because he made a number of gross scientific errors. He did not understand the basic physics of a bomb, never could come up with a bomb design, vastly overestimated how much U-235 (a processed form of uranium ideal for a bomb) was needed to achieve critical mass in the nuclear chain reaction, and did not understand that the Germans first would have to build a workable nuclear reactor before proceeding to construct a bomb. Above all, he failed to grasp that the bomb was as much an engineering problem as a scientific one. In the strict hierarchy of German science, no engineer would dare tell a scientist that a given concept was technologically wrong. Heisenberg, who was not an engineer, often had problems dealing with simple technical matters, and was not about to listen to any engineer; in German science, a physicist of Heisenberg's reputation did not lower himself to listen to engineers.

At least a year before the fall of Germany in 1945, Heisenberg realized the attempt to build a German atomic bomb had failed: Not only were he and his researchers hopelessly stuck, but also it was clear that Germany had lost the war. In early 1945, the testing equipment of "the virus house" was packed up and shipped, along with Heisenberg and his team, to a cave in the small eastern German town of Hechingen, to protect the project and the scientists from Allied bombing raids. Aware that the end was near, Heisenberg spent his time in the last days desultorily trying to get a crude atomic pile to work before the inevitable arrival of Allied troops; in his spare time, he played mournful dirges on an organ in the local church.

Rounded up with other German scientists and sent to England, Heisenberg and "the uranium club" were shocked one August day when they heard a radio report about the American atomic bomb dropped on Hiroshima. At first Heisenberg refused to believe it; he had surmised that the Allies almost certainly were working on a bomb (since 1940, all mention of nuclear physics had disappeared from American and British scientific journals, and a number of prominent American and British nuclear physicists he knew seemed to have dropped out of sight), but it was inconceivable that the Americans could have succeeded where the vaunted German science had not. "A very large explosive device, probably," he said dismissively, but subsequently realized that the Americans had in fact succeeded, and that Hiroshima had been destroyed by an atomic bomb.

"Well, we are all second-raters now," said one of the German atomic scientists, Carl Friedrich Weizecker. No one contradicted him, for they knew he was quite right: The sun of German science had now set. The sun of the new master of science, the United States, had risen dazzlingly bright in the white-hot flash of the explosion over Hiroshima. To be sure, that explosion, judged strictly in scientific terms, represented a triumph. But it also represented an entirely new order of moral quandaries for those involved in bringing it to life—and none more so than the man most responsible for its existence.

• • •

Leo Szilard, a physicist, was one of an extraordinary group of brilliant Hungarian Jews born around the turn of the twentieth century who would go on to make singular contributions to science. All of them—including the physicist Edward Teller and the mathematician John von Neumann—planned careers of pure science, with no thought of turning their talents to war. Szilard had followed the standard pre-Hitler path of higher science, studying at science's great temples of the University of Berlin and the University of Göttingen. In 1929 he became one of the scientific elite with groundbreaking studies in what is now called information theory. While in Berlin he became a close friend of Albert Einstein (together they patented a new refrigeration system that utilized electromagnetic waves, a design ultimately adapted for the liquid metal coolant in "fast breeder" nuclear reactors). Like Einstein, he was a committed pacifist who was constantly alert for any hint of totalitarianism that would threaten to convert pure science into an instrument of war. He watched the rise of the Nazis in Germany with growing alarm; by 1932, certain that Hitler eventually would come to power and that war was inevitable, he kept two suitcases packed in his room. The moment Hitler became chancellor of Germany in 1933, Szilard picked up his bags and fled to England. There, he went to work at a hospital laboratory, conducting research. One day, Szilard attended a lecture by the distinguished head of the Cavendish Laboratory, Ernest Rutherford, and heard him confidently say that although it was theoretically possible that someday man might release the tremendous inherent power in the atom, as a practical matter it would never be done. Szilard wondered about that sweeping assertion: As a scientist who had spent much of his graduate work in thermodynamics, he knew that the history of science was instructive—whatever had been deemed "impossible" always seemed to come to pass. He threw himself into the study of nuclear physics, and by his later account, one day was waiting at a traffic light when it hit him: Rutherford was right, but he hadn't gone far enough. True, the power locked inside the atom was tremendous and theoretically out of reach, but suppose there was a way to rupture the atom's wall and release all that energy? Conceivably, the power set off by such a release would be sufficient to set off still more ruptures in adjoining atoms, and still more ad infinitum, creating a chain of such events and releasing cataclysmic energy. After working out the mathematics, Szilard realized the enormous weapons potential of such a process, and decided he would never publish anything about it.

As he may have suspected, however, there was no way to keep so momentous a secret permanently squirreled away in his head. With growing alarm, he read of the Hahn-Strassman experiment, Niels Bohr's lecture, and the clear signs that several nations had embarked on nuclear weapons research. Most ominously, he learned that Germany had stopped the sale of uranium ore from Czechoslovakian mines now within its control, the unmistakable clue that the Germans had begun working on a weapon that would utilize the unstable uranium atom—an instability demonstrated by the experiment of Hahn and Strassman—to construct a chain reaction inside a bomb.

Szilard now found himself in the moral dilemma of his life. As a man of peace, the thought of a nuclear weapon, which threatened to become the greatest threat to mankind since the Black Plague, was anathema. Yet the possibility of a nuclear-armed Hitler was equally unthinkable. Finally he decided on a course of action that would cause him agony for the rest of his life: The United States must build a bomb first, by way of checkmating a German bomb. An American bomb, hopefully, would never be used, existing only as the ultimate deterrent.

As a scientist with a reputation only within the scientific com-

munity, Szilard had no hope of convincing the American government to commit the required resources to develop a nuclear
weapon. But he did have an invaluable asset: his friendship with Albert Einstein, the world's most famous scientist and a virtual icon in
the United States. In August 1939, as war was about to break out
in Europe, Szilard traveled to Einstein's summer cottage on the eastern end of Long Island and showed him the draft of a letter he
wanted Einstein to write to President Roosevelt. Basically, the letter
informed Roosevelt that nuclear weapons were probably feasible,
that war was inevitable, and that it was now up to the president to
decide what do. It took a follow-up letter a year later to get the
desired reaction: Roosevelt ordered the creation of what became
known as the Manhattan Project to build an American atomic
bomb, an effort in which the British Tube Alloys Project would be
melded.

Einstein shared Szilard's alarm about the possibility of a Nazi
atomic bomb, but in putting his signature to that letter, he became
science's most tragic figure. To his eternal dismay, the formula he
had devised for the purely scientific purpose of explaining certain
anomalies in Newtonian physics had been irrefutably proven in a
weapon of mass destruction. One can only imagine the pain he felt
to encounter the many popular accounts of the bomb's development
over the years that never failed to mention his famous mass-energy
formula as its basis. (Until his death in 1955, Einstein painstakingly and at great length answered letters from Japanese that accused him of being "the father of the atomic bomb," although he
was no such thing; nevertheless, he felt pained by the realization
that the formula contributed in no small way to the concept of such
a weapon.) And now, a few ink scratches on a piece of paper set in
motion a development program to produce a weapon that the pacifist Einstein could only find appalling.

Still, like Szilard, Einstein could justify it all on the irrefutable
point that world peace was threatened beyond measure by a Nazi
atomic bomb; therefore, doing the unthinkable in the face of such
a threat could be accommodated to his own deeply held beliefs
about the purity of science and the refutation of war as an instrument of politics. So did a number of other scientists who were
recruited into the Manhattan Project. Some of them, such as the
émigré scientists Hans Bethe and Edward Teller, had experienced

firsthand the nightmare of European totalitarianism. Others, such as the American physicist John Wheeler, would receive a direct reminder: His brother, serving as an infantryman in Europe, encountered the first evidence of Nazi genocide, which moved him to send his scientist brother a postcard that contained only two words: HURRY UP.

All of them shared a deep conviction that whatever their personal feelings, it was imperative that the United States, the only nation in the world with the technical, scientific, and economic capability to do so, develop an atomic bomb before Nazi Germany did. That they succeeded was due in part to their dedication and scientific talent, along with vast American resources (the project eventually would cost more than $2 billion in 1940s dollars and employ nearly 125,000 people), and, supremely, a scientific administrator of genius, J. Robert Oppenheimer.

A child prodigy who had completed his Harvard undergraduate education in three years, Oppenheimer earned his Ph.D. degree in physics in Göttingen, where he was among the glittering stars in a galaxy of scientific superstars. Later the head of the Radiation Laboratory at Berkeley, he personified the detached, near-ethereal pure scientist, a man who was surprised in 1932 when someone mentioned to him the 1929 stock market crash, an event of which he knew nothing, since he never read newspapers. Nevertheless, he had an acute sense of how the scientific mind worked, along with a shrewd eye for talent that he used to recruit the cream of the scientific community ("Oppie," as he was affectionately known in that community, could recruit in any of the eight languages in which he was fluent). What attracted these recruits to the project, aside from a sense of patriotism, was Oppenheimer's solemn promises that scientists would run this greatest of all scientific projects, that it would be run scientifically in the spirit of open-minded colloquia to which they were most accustomed, and that engineers would have equal status on the scientific team to work side by side with the scientists and determine the technological feasibility of any given scientific concept.

Those promises immediately put him at odds with the formidable figure of Colonel (later General) Leslie Groves, the military boss of the project, which was officially being funded by the army. Barely literate scientifically, he was an army engineer who had made his mark supervising construction of the Pentagon. Groves was a no-nonsense, hard-driving engineer with a volcanic temper (when angry, the veins in his neck would bulge dangerously) who had little regard for "pure" scientists who spoke in a language he didn't understand. "The biggest collection of crackpots and prima donnas you've ever seen," he told his men as they put together Los Alamos, a "science city" constructed in the wilds of New Mexico, where the scientific team would work and live.

Groves wanted to impose a military-style system on the project, including military ranks and uniforms for all the scientists (along with strict military discipline), and rigid compartmentalization for security reasons. Oppenheimer said no, and ultimately won, mainly because he told Groves that unless it was done his way, there would be no bomb. Reluctantly, Groves gave in and Oppenheimer ran what amounted to a free-flowing, university-type graduate seminar composed of various teams that attacked different parts of the problem, but met each other constantly to exchange ideas.

By late 1944 this system had made remarkable progress toward a workable bomb, at which point the Manhattan Project encountered its first serious crisis of conscience. American scientific intelligence teams that had followed troops into Germany found the scientific papers of Weizecker, a leading member of Heisenberg's "uranium club," which showed, beyond doubt, that the Germans had taken a scientific wrong turn; there was no possibility of a Nazi atomic bomb. Since the chief motivation of most of the scientists working on the project was to defeat that eventuality, they no longer saw much point in continuing work on an American bomb. They were not much taken with Groves's argument that the United States was still locked in a deadly struggle with Japan, a theater where the bomb might be used to save American lives.

Nevertheless, only one scientist, physicist Joseph Rotblat, resigned at that point (a political leftist, Rotblat claimed the only justification for the bomb now was to intimidate the Soviets). The rest handled the changed circumstances in different ways. Szilard and others of like mind, appalled at the idea that the bomb might actually be used to kill people, urgently advised that it be set off in a demonstration the Japanese would witness, sufficient to induce their surrender without the necessity of blowing up a city and God knew how many civilians. Others, such as the American Harold

Urey, argued that the bomb should be built as an ultimate weapon that would induce a postwar peace. And still others argued that the rest of the world scientific community eventually would learn how to build a bomb, so it was important that the United States have it.

Whatever the motivation, the project went forward, and on July 16, 1945, before a thousand witnesses, the world's first nuclear explosion, code-named Trinity, detonated in a New Mexico desert. Twenty-one days later, a bomb of similar power, about 10,000 tons of TNT, exploded over Hiroshima. After word of the successful bombing reached Los Alamos, Oppenheimer convened a meeting of the Manhattan Project scientists, which he opened by holding his clasped hands over his head, like a victorious prizefighter. The scientists sat in appalled silence, hardly able to believe that Oppenheimer, of all people, thought the deaths of some 127,000 human beings and the destruction of three-fifths of a Japanese city represented a cause to celebrate.

Oppenheimer's grisly exuberance would quickly fade, replaced by a growing sense of guilt. A sensitive man widely read in literature and philosophy, Oppenheimer had quoted the line "I am death, destroyer of worlds" from an ancient Hindu poem as the Trinity explosion lit up the New Mexico desert, and now, as the full import of what he had done sank in, he felt a sense of betrayal: He had betrayed the tenets of what science was supposed to be. His mood was not improved in 1946 upon meeting President Truman in his newly appointed capacity as head of the General Advisory Committee, created to advise the president on what the United States should do with its nuclear weapon. Truman, a hardheaded and practical politician, found it hard to communicate with Oppenheimer, who talked of nuclear scientists having "bloody hands." He brushed aside such philosophical ruminations to get to the crux of his major concern at the moment: When did Oppenheimer think the Soviet Union would acquire its own atomic bomb? Oppenheimer replied cautiously, noting that, contrary to popular opinion, nuclear energy wasn't that much of a secret, that the Soviets had competent nuclear physicists, and that with sufficient expenditure of resources, they probably would build a bomb fairly soon. He had not even finished when Truman interrupted: "I'll tell you when. Never."

Oppenheimer hardly knew what to say: Did the president of the United States actually think that the atomic bomb was the result of some sort of magic formula that could be kept secret forever? Worse, in Oppenheimer's view, was Truman's broad hints that he was about to approve the next, fateful, step in nuclear weapons: development of an American hydrogen bomb, a weapon that would use the fusion, rather than the fission, of atoms to achieve even greater explosive power. One of Oppenheimer's scientists in the Manhattan Project, Edward Teller, had been working on what he called "the super" for some time, and now had the mathematics worked out that proved such a bomb was feasible.

Originally, Oppenheimer had called the concept "technically sweet," but now thought it was a terrible idea. In his view, it would simply create an even more terrible weapon of mass destruction that would stimulate the Soviet Union to match it, setting off an arms race that could only end in Armageddon. Oppenheimer lobbied against developing the thermonuclear weapon, but quickly learned that in the new Cold War, scientists were expected to conform. Estranged from the military establishment because of his opposition to developing thermonuclear weapons, he found himself squeezed out of the decisionmaking process, climaxed by his removal as head of the General Advisory Committee on the grounds that he was a "security risk." (Oppenheimer's wife had been a member of the U.S. Communist Party, as were several other relatives, and he had prevaricated about an attempt by his brother-in-law, an asset of Soviet intelligence, to recruit him for espionage.)

To Oppenheimer, all this represented an eerie echo of events some three hundred years before, when clergy and laity in Galileo's time were hopelessly ignorant about the "new science," and felt threatened by a man they did not really understand. Now, neither the military nor the general public seemed to have any grasp of just what nuclear weapons were all about. General George C. Marshall had returned a report Oppenheimer had written concerning how the bomb worked with the comment that he found it "impenetrable," and the popular press was full of sensational stories about cartoon-like mad scientists who had found the fabled death ray of science fiction—an ignorance aided in no small part by the tight veil of secrecy that surrounded nuclear weapons, even including basic nuclear physics.

Oppenheimer was clearly out of step with both the government and the popular mood, which was to maintain the American monopoly at any cost. The justification was the Soviet Union, which, everyone seemed convinced, had embarked on a hellbent race to catch up. On that point, at least, everybody was right.

• • •

"Russia has suffered because of her backwardness," Joseph Stalin was fond of lecturing his minions. "We will never be backward again." Stalin was correct in his diagnosis of Russia's ills: Since the time of Peter the Great's leap forward in the eighteenth century, Russian science and technology had fallen many decades behind everybody else's. The decline was dramatically exposed in both the Russo-Japanese War of 1905, when Russia was defeated by a nation a fraction of its size, and World War I, when the world's largest army was shattered in a catastrophic defeat in which Russia itself collapsed. These disasters, Stalin concluded, stemmed from Russia's inferiority in science and technology, an inferiority that kept it in a state of perpetual danger from more scientifically advanced states—most dangerously, Germany. But his cure, a massive scientific and technological bootstrap to lift the Soviet Union from backwardness, would, in the end, destroy the very nation he was trying to save. And that tragically ironic course of events had everything to do with Joseph Stalin himself.

Xenophobic, paranoid, and totally ruthless, Stalin had a schooling barely beyond fifth-grade level, and like his greatest enemy, Hitler, was a scientific illiterate. All Stalin knew was that the Soviet Union at its birth appeared to be hopelessly behind in every measure of national power; in 1921, when Stalin began his rise to eventual domination over the Soviet Union, its iron production was only a fifth of the 1913 level, and coal production less than 3 percent of that year's level. Most of the nation's scientists and technicians had fled the country during the civil war. The situation was so bad that Leon Trotsky, architect of the Red Army, had to hire foreign scientists, technicians, and engineers to keep the army operating during the civil war.

In 1929 the Soviet Union's first Five-Year Plan strongly emphasized science, calling for the building of science academies, research institutes, and the recruitment of "aspirants" (prospective scientists). The plan set a goal of a thousand scientists by 1930; by 1935, the goal was six thousand. In 1939, on the eve of the outbreak of

World War II, some eighteen hundred science institutes were in operation, along with 570 institutes for "industrial research," a euphemism for military research and development centers. But Stalin, who by that time had seized the reins of power, made it clear what kind of science he wanted in all those institutes: any science that would improve the Soviet military. In 1929, when he thought the USSR Academy of Sciences, the overall coordinating agency for Soviet science, was spending too much of its time on pure science without much practical use, its leaders were denounced for running "a center for counterrevolutionary activity." Three of them were shipped off to slave labor camps in Siberia, replaced by Stalinist *apparatchiks;* the rest fell dutifully into line, making sure no science was conducted in the Soviet Union unless it was directed toward practical (almost always military) application.

Gradually even this power was removed from the academy, which became a virtual rubber stamp for Stalin, who devolved upon himself the task of final arbiter of what kind of science the Soviet government would support. A greater disaster for Soviet science cannot be imagined. Obsessed with bigness, Stalin sank huge sums into vast scientific and technical extravaganzas that simply didn't work—such as the White Sea Canal, which cost the lives of two hundred thousand slave laborers (and which turned out to be useless), and the world's biggest bomber, a huge, six-engined behemoth Stalin liked to show off by having hundreds of paratroops drop off its wings like so many flies (but the bomber was so slow that Soviet pilots joked someone could bring it down with a well-aimed rock). He personally interfered in virtually every scientific field, most disastrously in genetics, where he decided that a "peasant scientist" named T. D. Lysenko (a fraud who believed that Mendelian genetics was bunk and that improved crops could be obtained by genetic manipulation, not over a period of years, as modern genetic science had concluded, but immediately) be put in charge of all Soviet agricultural genetics. Lysenko proceeded to destroy Soviet agriculture, a disaster from which it never recovered. (While Soviet agriculture was collapsing, American plant geneticists, among other accomplishments, achieved the hybridization of corn, increasing crop yields 80 percent; by 1945, American agricultural production had more than doubled from its 1935 levels. By contrast, in 1966, a desperate Soviet Union, facing widespread food shortages, began importing grain—an astonishing admission of failure in a nation once known as the "breadbasket of Europe" and that in 1913 could export 40 percent of its grain because it produced so much. The failure was one of the main reasons for the eventual collapse of the Soviet Union.)

Given this range of scientific disasters, it would seem a wonder that the Soviet Union managed the considerable scientific and technical feat of developing nuclear weapons just four years after Hiroshima. The reason, again, was Stalin. Although he liked to interfere in virtually every scientific field, he left alone certain sciences he felt critically important to building his military power, especially physics, metallurgy, and chemistry. Indeed, scientists in those fields became the only aristocracy in the Soviet Union—paid lavishly, given privileges ordinary Soviet citizens had no access to, and working under pleasant conditions in one of the "science cities" Stalin built all over the Soviet Union, self-contained metropolises where scientists worked and lived in a luxury ordinary Soviet citizens could only dream of achieving.

Beginning in 1939, Stalin's most pampered scientists were the nuclear physicists, the men he was convinced would provide him with an atomic bomb. His particular favorite was Igor Kurchatov, who headed the Soviet nuclear weapons development program. A slavish Stalin acolyte who was known as "the beard" for the lengthy beard in the shape of a flange that extended below his chest, Kurchatov, despite his education, was pure peasant, the kind of man who urinated in laboratory sinks. He had begun his scientific career as a "pure" scientist investigating the dynamics of the atom (he and a team of physicists built the world's first particle accelerator in the 1930s), but when Stalin assigned him the task of developing an atomic bomb (and doing it as quickly as possible), Kurchatov eagerly went to work. To do so required a certain moral myopia—the thousands of slave laborers mobilized to mine uranium, the scientists purged on Stalinist suspicions they were insufficiently enthusiastic about developing a weapon of mass destruction, the ruthless diversion of resources from an already hard-pressed civilian economy (Kurchatov and his fellow scientists ate very well, while the average Soviet citizen suffered under severe food rationing).

Not all Soviet nuclear physicists shared Kurchatov's unquestioning devotion to building a bomb, nor his moral blindness. Peter Kapitsa, among the most brilliant Soviet physicists, had spent several years working at the Cavendish Laboratory in England under Ernest Rutherford on atomic research, but was summoned home for a more urgent task when Stalin decreed the building of a Soviet atomic bomb. Like some of his American counterparts, Kapitsa faced a crisis of conscience: He did not want to work on a weapon of mass destruction, yet the possibility of such a weapon in the hands of Nazi Germany, the Soviet Union's greatest threat, was unthinkable. Accordingly, Kapitsa trudged sadly home, leaving behind a poignant letter to an English friend in which he lamented the passing of nuclear science's innocent age: "What a wonderful time it was then at the Cavendish. The science belonged to the scientists and not the politicians."

Back home in the Soviet Union, Kapitsa quickly learned that nuclear science turned to war indeed belonged to the politicians. He was appalled at the platoons of NKVD guards who watched every movement of the scientists and the sheer terror in which the scientists worked, fearful that even the slightest slip-up would land them in the Gulag as "saboteurs" or "imperialist wreckers." He was especially shaken by the numbers of scientists who had been dragged away to prison on the whim of Stalinist suspicions about their loyalty and devotion to nuclear weapons. Many of the victims were Jews; Stalin had become convinced that Jews were part of a vast international conspiracy. Among those caught up in this Stalinist paranoia was Lev Landau, among the more brilliant Soviet physicists, thrown into prison as a "German spy." Kurchatov had said nothing about this outrage, but Kapitsa personally intervened, demanding that Landau be released forthwith—otherwise, Kapitsa said, he would refuse to do any work on a Soviet atomic bomb. It is a measure of Stalin's eagerness to get a bomb that he acquiesced; Landau was released, and Kapitsa devoted his talents to nuclear weapons.

By December 1941 the Soviet program had made remarkable forward strides, despite the German invasion. By the day of Hiroshima, the program was well along, further spurred when an anxious Stalin ordered Kurchatov to get a bomb together "before the Americans blow us up." One last remaining technological barrier remained: How to set off the explosion that in turn ignited the nuclear reaction. The problem was finally solved in 1945 when Klaus Fuchs, an émigré German physicist working in the Manhattan Project who had been recruited by Soviet intelligence, revealed that the Americans used a unique implosion process. With that last barrier cleared away, the Soviet program went forward, and on August 29, 1949, tested its first atomic bomb. Kapitsa pointedly was not among the Soviet nuclear scientists who watched the explosion; four years before, he had refused to do any further work on nuclear weapons. Surprisingly, Stalin didn't order him hauled off to the Gulag, although Kapitsa, stripped of all rank and privileges, was put under de facto house arrest for the rest of his life.

• • •

And so the union of science and war had brought mankind to the edge of the precipice. Men such as Oppenheimer, Szilard, and Kapitsa were sickened over what had happened and how much they had contributed to the course of events. They fell into a deep pessimism, convinced that nothing they said or did would make the slightest difference; nobody seemed to be listening. Discouraged, they retreated into private worlds: Szilard gave up physics and began research work in microbiology, while Kapitsa gave up science altogether, and Oppenheimer stopped working on applied physics. They watched as the new science of ultimate destruction solidified the centuries-old belief that science now held the keys to power. But in this new dogma, scientists were expected to adhere to the political institutions of the state, the godhead that would pamper and shelter them—as long as they conformed. Scientists quickly discovered that in this atmosphere, the long-cherished independence of science was rapidly evaporating—witness the tragicomic ordeal of the prominent American physicist Edward U. Condon, who in 1948 found himself before a Loyalty Board hearing that regarded his boyhood job of delivering Socialist newspapers with grave suspicion.

"Dr. Condon," a member of the board began with somber seriousness, "it says here that you have been at the forefront of a revolutionary movement in physics called quantum mechanics. It strikes this hearing that if you could be at the forefront of one revolutionary movement . . . you could be at the forefront of another." In reply, Condon just as solemnly held up his right hand and swore belief in a number of scientific revolutions, including Archimedian mechanics, Galilean motion, and Newtonian physics.

Oppenheimer found the incident amusing, although as a scientist who had undergone a similar, but more trying, ordeal, he worried about the state's growing obsession with subjecting all scientists to a political litmus test. He contemptuously called the scientists who conformed to that process "experimentalists." Perhaps, however, his contempt was misplaced: In a 1951 lecture at Berkeley, he held up a copy of Beethoven's Quartet in C-sharp Minor, among his favorite pieces of music, as an example of the kind of refinement he now argued was supremely important, as opposed to the crasser concerns of his cruder cousins in the scientific community. Apparently he was unaware that Beethoven, in a note to his publisher when the work was finished, instructed: "It must be dedicated to Lieutenant General Field Marshal von Sutterheim."

9

The Age of Doom

*When the storm rages and the state is threatened
by shipwreck, one can do nothing more noble
than to lower the anchor of our peaceful studies
into the ground of eternity.*

—JOHANNES KEPLER, 1629

At a range of nearly 200 miles, the radar screens came alive with a series of fast-moving, blue-green blips that morning of June 10, 1972, quickly identified as a formation of U.S. Air Force F-4 Phantom fighter-bombers moving on a heading directly toward the Lang Chi hydroelectric facility in North Vietnam. At first, the Soviet military advisers overseeing the batteries of surface-to-air (SAM) missiles and nests of radar-directed antiaircraft guns defending Lang Chi thought the Americans must have made a mistake; in eight years of bombing North Vietnam, the Americans had never attacked the facility, despite the fact that it supplied 75 percent of North Vietnam's electrical power.

But as the seconds ticked by, it was clear there was no mistake: The Americans clearly intended to attack Lang Chi. An alarm sounded and the missile and antiaircraft gun crews readied for action. Superbly trained by their Soviet advisers and armed with the most technologically advanced weapons in the Soviet inventory, the crews went to work confidently, secure in the knowledge that the extensive air defense system built by the USSR in North Vietnam had taken a terrible toll on American aircraft during the past eight years, the major reason why the American bombing of North Vietnam was ineffective. As the Soviets had taught their North Vietnamese pupils, the Americans had magnificent high-performance

jet warplanes, but were forced to throw away that advantage by relying on "dumb" (meaning gravity-guided) bombs, pretty much the
same kind of "city buster" bombs used in World War II. To achieve
bombing accuracy, the planes had to slow down their speed and
swoop in low—at which point they were sitting ducks (or, alternatively, drop bombs from higher altitudes, almost guaranteeing inac-
curacy).

And now the Americans were headed into the teeth of the bristling defenses of Lang Chi, which the Soviets and the North Vietnamese had eight years to turn into the most heavily defended place
on earth: multiple batteries of SAMs and dozens of radar-directed
antiaircraft guns. The Soviet experts had told Moscow that Lang Chi
was impervious to any American air attack, and it was difficult to
argue with that claim; some of the Soviet advisers joked that the
area was so weighted down with missiles, radars, and guns, it was
about to sink into the Red River delta.

There was an even more important reason for the Soviet confidence in Lang Chi's invulnerability: The Americans couldn't bomb
it without politically unacceptable heavy civilian casualties. The
Lang Chi complex, built by the USSR as a showpiece of Soviet technology, included a large dam holding back billions of gallons of
water that if released, would drown twenty-three thousand civilians
living in a densely populated area nearby. Accordingly, Lang Chi had
been off-limits to American bombers since 1964, since neither air
force nor navy planners could guarantee that the dam would not be
hit in the process of bombing the main power station.

But in 1972, Lang Chi came under attack for the first time, and
what happened there that June morning would mark still another
major revolution in the conduct of war. And like all previous such
revolutions, it also represented a fusion of science and technology
whose ripple effects would spread very wide.

• • •

As the American F-4 Phantoms came within SAM range, the North
Vietnamese prepared to fire. Hardly able to believe that the Americans would foolishly attack so heavily defended a target (or hit it
without the politically unacceptable "collateral damage" of heavy
civilian casualties), the radar operators, rocket technicians, and gunners followed the blips on the radar screens with confidence that

they would exact a heavy toll. By now they were familiar with the American bombing tactics: Several "Wild Weasels," jets with special electronic detection gear, would come in first, deliberately exposing themselves to SAM and antiaircraft fire to spot, and then hit, rocket and gun sites with air-to-ground missiles. Then would come the bomb-laden Phantoms, twisting and turning as they evaded SAMs and gunfire to roar in as low as possible and guide their bombs to the target.

But on that morning in 1972, the North Vietnamese and the Soviets quickly realized that everything had changed. The radar operators reported that their scopes were being blinded by bursts of mysterious, high-energy electronic interference from some of the American planes; other planes had launched missiles that followed the radar beams back to their sources, blowing up the screens and computers. Then the real shock occurred: Instead of an entire flight of planes swooping in low to drop their bombs, there were only four planes, and none of them came in low. Instead, they stayed at high altitude and dropped a dozen bombs. All twelve fell unerringly into a 50-foot-by-100-foot section of the roof of the main power plant and blew up inside; the building virtually disappeared in a pile of rubble and dust. There was not even a scratch on the facility's dam or spillway, and no civilian casualties.

Frantic reports by the Soviet advisers back to Moscow underscored the militarily disastrous turn of events: The Americans had come up with new "miracle weapons" that blinded even the most advanced radar, blew up radar stations with missiles that somehow honed in on their electronic signals, and placed bombs unerringly into even small targets from high altitude in all weather and through cloud cover, the technological equivalent of throwing a baseball into a glass jar from 200 feet away. Over the next six months, the news went from bad to worse: An all-out air assault by the Americans took the country's infrastructure apart, piece by piece. It had been set in motion by President Nixon, who, determined to end the deadlock in the Vietnam peace talks, had decided on a military solution: The full might of American air power was unleashed, without restriction, to bomb North Vietnam into submission. Nixon was confident that this bombing campaign—unlike the others during the previous eight years—would work, for he was now armed with new, deadly science and technology that made the ordinary jet fighter or bomber a truly formidable weapon: laser-guided bombs, air-to-ground missiles with "seeker" electronics that "rode" radar beams back to their transmitters, air-launched cruise missiles with television-guided warheads, and electronic jamming and countermeasures that scrambled radar signals.

To the shock of the Soviets, by Christmas of 1972, their state-of-the-art air defense system, installed at a cost of billions of rubles to protect North Vietnam, was in shambles. And even the parts that still worked were totally ineffective: on Christmas Day, a flight of F-4s put sixteen laser-guided bombs through the roof vents of North Vietnam's main SAM assembly plant in Hanoi, blowing it up in a huge explosion that broke windows for miles around. The Phantom pilots couldn't even see their target, but hit it precisely from an altitude of 22,000 feet through solid overcast. Batteries of SAMs fired a total of forty-eight missiles, but with their radar guidance systems electronically blinded, none of the missiles came even close. Within weeks, the diplomatic impasse was broken, and America's war in Vietnam was over.

• • •

The science and technology may have been different, but the dramatic turn of events in Vietnam in 1972 followed the pattern set long before in other scientific-technical revolutions that transformed war—a perceived military need, a harnessing of scientific and technical resources to achieve it, and a resulting scientific leap forward. From the introduction of the wheeled chariot in the ancient Near East millennia ago, transforming the civilized world, to the laser-guided bombs that crashed through the roof of a hydroelectric powerhouse in the modern Far East, revolutionizing modern war, science has been the indispensable engine that drives change. But the important difference between the time of the chariot and the laser-guided bomb lies not only in the accelerated pace of change and the increasing power of weapons, but also in the formal institutionalization of scientific change as a weapon of war.

That process was born the day Dr. Vannevar Bush walked into President Roosevelt's office and proposed the creation of an institutional structure to marshal American scientific resources in a coordinated effort to win World War II. That effort turned out to be spectacularly successful; it was American superiority in science and technology, combined with an industrial capacity capable of producing gargantuan amounts of technologically advanced weapons, that was largely responsible for the Allied triumph. The success led Bush in 1945 to propose making the wartime Office of Research and Development permanent, with the aim of harnessing science to keep the American scientific-technical lead. Eventually, following a struggle concerning how much control the scientific establishment would have over a wartime structure now converted into a weapon of the Cold War, in 1950 the National Science Foundation was established to coordinate federally funded scientific research. By 1956 it had disbursed $3.45 billion, more than 80 percent of it military-related. The launching of Sputnik a year later set off something of a panic in the American political establishment; as an alarmed Senator Lester Hill of Alabama phrased it, "The Soviet Union . . . is challenging our America in the application of science to technology." The result was the National Defense Education Act, created in the belief that only a massive federal investment in the basic sciences would lead to scientific and technical superiority over the Soviet Union. To a large extent it stemmed from a classic American prejudice, confusing a high standard of living with scientific superiority; Americans assumed that cars, air conditioning, and frozen food meant they were scientifically far ahead of the Soviet Union, where such consumer products were virtually nonexistent. How such a backward society somehow managed to build rockets powerful enough to launch a 184.3-pound satellite into earth orbit was a severe shock to American public opinion. Although the general public feared that the Soviet Union had surged ahead, scientists understood perfectly well what had happened: The Soviets had concentrated their resources into one particular scientific discipline—in this case, aerodynamics—to produce a particular result, the same kind of result the Manhattan Project was designed to produce.

But the important consequence of Sputnik to American science was political: Thanks in no small part to the lobbying of the scientific establishment, Congress approved the National Defense Education Act, created to upgrade science education and to build a vast pool of scientists who would reestablish American scientific dominance. It allocated the kind of money science had never seen before, even in World War II: $100 million of student loans for science

majors, $300 million for science equipment, and $1 billion worth of
fellowships for graduate students in science. Within five years, the
act tripled scientific research and education and created a large
reservoir of scientists and technical specialists. Nearly forty-four
years later, the federal government was lavishing more than $75 bil-
lion annually on this reservoir of scientific talent, more than $40
billion of it devoted to military purposes.

This political turn of events, born of a near-panic that the United
States had fallen behind, formed the foundation for what would become an American scientific and technical behemoth, one unparal-
leled in all history. Largely driven by a conviction that military
superiority is essential for American national security, scientific innovation and its direct descendant, application of new technology,
are together responsible for more than half of the growth in the
American gross national product since World War II. Technical and
scientific advances generated by deep federal investment in science
have not only created the most powerful military machine in world
history, they have also created an economic juggernaut; the steady
flow of discoveries out of research laboratories and universities has
been fed directly into the industries that drive the economy.

This scientific cornucopia, however, came with a price, one that
has caused no end of trouble for science: secrecy, the inevitable consequence of government's deep reach into science. It was most obvious in nuclear physics, where in 1946 the newly enacted Atomic
Energy Act gave the government, via the Atomic Energy Commis-
sion (AEC), total monopoly over all atomic energy matters, both
military and civilian. This unprecedented control over a scientific
discipline often blurred the thin line between legitimate national
security concerns and the military's desire to control all scientific and technical information, reaching such absurd lengths as
the AEC's pressuring *Scientific American* to quash an article by the
Manhattan Project physicist Hans Bethe on the basic science of
thermonuclear weapons (as if Soviet nuclear physicists knew nothing of thermonuclear reactions). There was also the contentious
Progressive magazine case, in which the government moved in
court to force the magazine not to publish an article on the same
subject. A number of nuclear physicists wrote letters to the AEC,
noting that the government's action was pointless in the *Progres-*

sive case in as much as the scientific details in the article were contained in an encyclopedia article by Edward Teller some years before (the government promptly classified the letters "secret").

In an atmosphere of traditional scientific openness and the American system of accountability, these security classification imbroglios created no end of friction between science and government, along with a growing public uneasiness that there was something very significant going on out there and they weren't being sufficiently informed about it. No American citizen who saw the newsreels of the first test of the hydrogen bomb on November 1, 1952, could doubt that science had brought to life something awesome, dangerous, frightening—and perhaps uncontrollable. Even in the flat gray and white tones of a newsreel, the effect was stupendous: a fireball 3 miles wide from an explosive force 100 times the power of the Hiroshima bomb, with a mushroom cloud that rose 57,000 feet into the air in 90 seconds and eventually spread 1,000 miles in width. The entire island of Ewgelob disappeared, leaving an underwater crater 6,240 feet wide and 164 feet deep.

However much this superweapon was shrouded in secrecy, there was no hiding its cataclysmic power from the public it was supposedly protecting. Anyone who had paid attention in their high school science classes understood that the weapon captured the thermonuclear process of the sun, fusing hydrogen atoms at temperatures approaching 50 million degrees Centigrade (achieved by igniting a fission bomb as "trigger"), ejecting high-speed neutrons in the process, and setting off a massive amount of explosive power as mass was converted into energy.

They understood much less about the complex motives of the scientists behind it, such as Edward Teller, who had begun his scientific career devoted to studying how the sun worked, and wound up nearly obsessed with building weapons of mass destruction. But they did hear enough to wonder if scientists in the Teller mode—a composite of them would form the "Dr. Strangelove" character and contribute the word "Strangelovian" to the English language—had begun to lose their grip on reality. Teller made a series of highly publicized proposals in the 1950s for "peaceful" uses of thermonuclear weapons, including setting off a hundred hydrogen bombs to dig out a canal across Israel, and another series of thermonuclear explosions to carve a canal across Greece. (This latter idea, proposed directly to

Queen Frederika of Greece, prompted her to fix him with a withering stare and reply, "Thank you, Dr. Teller, but Greece has enough quaint ruins already.")

The growing public uneasiness with some of the more Frankensteinian aspects of what modern science had provided extended into the scientific community itself. Groups of concerned scientists began to coalesce around some of the scientific issues that were causing increasing disquiet among scientists and nonscientists alike, among them the world-threatening power of nuclear weapons, the dangers of nuclear fallout, the health threat from such new powerful agricultural chemicals as DDT, and the concern over carcinogenic food additives. All these issues represented consequences of science developed for war (e.g., DDT was developed to eliminate malaria-bearing mosquitos endangering troops operating in jungle areas), and there was a gathering sense that much of the science had grown out of control. For the first time, the legendary "mad scientist" of comic strip and film—the spade-bearded, bushy-haired, maniacal-looking man in white coat prone to utter "With this, I shall be master of the world!" in a crude German accent as he brandished a test tube filled with smoking liquid—became, for many people, the dominant image of modern science.

It was an image that concerned the scientific establishment, which liked to point out that science hadn't invented industrial pollution or the satanic mills of the industrial revolution; these were the consequences of technology, applied science. True to a certain extent, but the fact is that modern science and technology have been generally regarded as synonymous, largely because so much of modern science was first developed for war, then left for the technologists to develop for other purposes. The scientists who developed the aerosol spray can in World War II to combat mosquitos argued that they had no idea that their invention would lead to a consumer revolution—nor that the chemical miracle they devised would come to represent a serious threat to the planet's ozone layer. Still, a number of scientists came to believe that the argument was too facile, that modern science and scientists bore the ultimate responsibility for what they had brought forth. Further, the problem really centered on science's tradition of ceding control of science to government and industry. Science had to become more involved in deciding how science would be used. This idea took its firmest root, unlikely enough, in the Soviet Union, where one of its most brilliant scientists underwent a profound crisis of conscience.

• • •

Like most other nuclear physicists, Andrei Sakharov began his career as a pure scientist, in his case the study of cosmic rays. His 1947 doctoral dissertation on that subject attracted the strong interest of Igor Tamm, among the nuclear physicists enlisted for the Soviet Union's nuclear weapons program, since cosmic rays play a critical role in thermonuclear fusion. Tamm recruited the brilliant young physicist into the program in 1948, and only two years later, they jointly obtained a controlled thermonuclear reaction, the essential basis for a thermonuclear weapon. Sakharov went on to develop the Soviet Union's hydrogen bomb, first tested in 1953, for which he received the first of his three Orders of Hero of Socialist Labor.

As a superstar of Soviet science, Sakharov lived a life of privilege—expensive apartment; shopping in "privileged" stores reserved for the Soviet elite; and work in a "science city," a cocoon isolated from the problems faced by ordinary Soviet citizens. Soviet science had been thoroughly co-opted by the state, and content with their privileged status, Soviet scientists tended to ignore the implications of their work and existence. But Sakharov, despite his status, began to conclude that science in general—and Soviet science in particular—had wandered into moral oblivion. He had seen the long lines of wretched slave laborers herded through the streets of his science city like so many cattle by guards with whips and dogs, on their way to work in the uranium production factories (few would survive). And he had heard of the disaster in the Soviet rocket program, where one day an explosion on a launching pad killed two hundred scientists and technicians. "Clean up the mess and get back to work," Lavrenty Beria, the Soviet secret police chief in charge of the program ordered, as if the dead men were just so many pieces of trash.

Sakharov's uneasiness increased as he realized that Soviet science had no interest in the consequences of what it was contributing to the military power of the Soviet Union, a fault he came to believe was shared by American science. Among the first scientists to understand the danger of nuclear fallout, Sakharov encountered only disinterest when he tried to alert his fellow scientists to the threat.

Worse, he encountered equal disinterest when he raised the subject with the head of the Soviet Strategic Rocket Forces, the organization entrusted with the task of launching thermonuclear weapons to kill millions of human beings. Sakharov told the bemedaled general of vast clouds of radioactive fallout that would spread for thousands of miles, rendering much of the world uninhabitable for millennia, and the danger to the human race from the radioactive fallout pumped into the air by atmospheric nuclear weapons tests. "Such matters are no business of scientists," the general said dismissively. "These are matters for the Central Committee."

Sakharov later described this moment as the epiphany in his life; from then on, he became a relentless crusader to ban atmospheric testing and dismantle nuclear arsenals. A brave stance for a Soviet scientist, and it would prove costly to him: revocation of his privileged status, relentless harassment by the state, life as a virtual pariah in his own country—and the Nobel Prize for Peace. Sakharov felt he had proven that science had a conscience and that scientists could influence how science would be used. That was true to a limited extent, for Sakharov's concept of influence was restricted to his own scientific discipline, nuclear physics, where there was a direct line between concept (nuclear weapons were born strictly for military purposes) and use (no other purpose for nuclear weapons exists except for destruction). But the lines were considerably less clear in many other scientific disciplines, where military-related research could lead to unforeseen consequences very difficult to control.

Among the more illuminating examples is the laser, which grew out of wartime research to replace high-quality German glass, the essential component in fire control instruments. The research to develop a superior American glass led to a number of spin-offs, including fiberglass, but it also resulted in some original thinking: Could new forms of glass be used to transmit radio waves underwater to track submarines? Eventually that led to the discovery of vibrating molecules in certain kinds of glass (ruby glass turned out to be the best) to release microwave radiation. From there, the next step was to "dope" certain forms of glass with neodymium, which created a powerful beam in a process known as "light amplification by stimulated emission of radiation" (or, more commonly, laser).

There was also the transistor, an unperfected advance originally intended to improve telephone service that languished in a scientific backwater until World War II, when the critical need for improved electronics (primarily to replace the bulky vacuum tube) brought in a huge amount of money and research talent that finally produced a breakthrough: Adding certain impurities to atoms in semiconductor crystals caused "holes" into which electrical fields could be introduced, resulting in unidirectional current much faster than that produced in vacuum tubes (and they lasted much longer, too).

None of the scientists involved in either the effort to devise better glass or improved electronic communications could possibly have foreseen the sheer scale of the military consequences of their work—miniaturization of components to make possible hydra-headed nuclear warheads, guidance systems for cruise missiles, beams for the pinpoint accuracy of "smart bombs," electronic countermeasures. These developments, which came about because of the perceived need for advanced military technology and the government's willingness to fund the effort to invent them, represent the Law of Unintended Consequences. Unlike nuclear physicists who from the first moment understood they were working on a weapon to blow things up, no scientist involved in the effort to stimulate molecules in a piece of glass could have realized that one day that effort would allow some F-4 Phantoms to drop bombs through the roof of the Lang Chi hydroelectric plant.

But there is an even greater (and in many ways most curious) example of how that law, in the context of war, can result in momentous scientific and technical change. It involves a machine that is responsible for the ongoing scientific revolution known as the Information Age.

• • •

The science of mathematics was drawn into war because only that discipline promised a way into unlocking the armor that guarded modern cipher machines. Before the war, mathematicians could not have guessed that they would become warriors (what, after all, is the possible military use of Boolean algebra?), and after the war, they reasonably assumed that their military work was over. Actually, it had only begun, for although one war was over, another, the Cold War, had broken out—and one of its battlefields was the ether, where millions of electronic signals flowed. Those signals represented essential electronic tendrils of modern war, and they were tendrils the modern military wanted to read.

Privately, officials of the National Security Agency (NSA), the main U.S. communications intelligence agency, like to boast that their organization is almost entirely responsible for the Information Age. It is only a slight exaggeration, for indeed the effort to decipher everybody else's electronic signals underlies all modern computer technology. Largely, that technology stemmed from a frantic, covert race that began with the mathematician Alan Turing's *bombe:* bigger, faster, and better cipher machines to generate an even greater number of mathematical possibilities with each keystroke; faster and more powerful computers to keep up; and on and on, in an upward spiral of complexity. By the 1950s, the NSA had begun hiring battalions of mathematicians to devise unbreakable codes (while eventually succeeding in cracking the codes of 104 countries), along with electronic engineers to design new computing machines to handle the dizzying advances in complexity. The silent war between codemakers and codebreakers was almost entirely waged on a battleground of mathematics, as mathematicians delved deeply into the more esoteric aspects of their discipline to enlist some very unlikely recruits, including the ancient Greek mathematician Euclid.

At first glance, Euclid's speculations about prime numbers (divisible only by 1 or itself) and composite numbers (numbers greater than 1 that must have divisors other than 1 and itself) seem of interest only to "pure" mathematicians. But such numbers turned out to have immense implications in the world of codes and ciphers. That's because Euclid postulated that there are an infinite number of primes, a postulate not yet completely proven because even the most powerful modern computers have been able to find prime numbers only up to 65,050 digits, not even close to what is believed the ultimate total. Creating ciphers with prime numbers opens so many mathematical possibilities, verging on the infinite, that they are theoretically unbreakable. The answer has been ever more powerful computers to handle the huge volume of possibilities—such as the first generation of Cray supercomputers, which could handle 70 quadrillion calculations in five seconds.

While the NSA was funding rapid improvements in computers, the military, building on its experience with the first crude computers in World War II, also provided a major spur to the computer rev-

olution. The military had realized that in modern war, it was encountering data processing problems only better computers could
solve. The major need came in 1949, with the construction of the
National Defense Command Center inside a Colorado mountain.
The center, mandated to keep track of everything anywhere near
American airspace, and provide at least four hours' advance warning
of approaching Soviet bombers, was linked to a chain of early-
warning radar stations in northern Canada. The daunting task of
tracking, analyzing, and coordinating all that information meant
considerable computing power—achieved by the convergence of
transistors, enhanced power, and miniaturization with solid-state
components. The convergence came about because of lavish military research and development funds, and the result was the first
computer with a family resemblance to the machines on desktops
today. Equipped with new machine language that allowed the computer to "read" the information fed into it and software—essentially
coded instructions that instructed the computer to perform certain
tasks in a certain sequence—the descendant of the codebreaker Alan
Turing's *bombe* was now ready to revolutionize the world.

It is difficult to imagine this remarkable scientific and technical
development coming to fruition in so short a time without the military impetus that brought it into being. Conceivably, without war
someone eventually may have developed Turing's original prewar
conception into a viable computer, but there is no doubt it would
have taken many more decades to achieve. The simple fact is that
in the beginning there was no need for computers, except for military requirements. That requirement assumed a civilian aspect only
after an airline executive, seated beside an IBM engineer aboard a
flight one day in 1953, bewailed the burden his company bore trying
to handle flight reservations. After the IBM engineer told him about
a wonderful machine called a computer that was helping the air
force keep track of all that stuff in the air, the airline executive realized he had found the answer to his problem. In short order, he had
IBM build him a version of the machine to manage the task of coordinating flight schedules, reservations, and all the other bits of
information necessary to sell tickets. Word of how well the new machine worked soon spread throughout the business community, and
in short order, the civilian computer revolution was born.

The rest of the computer story has become the stuff of legend—

the college dropout named Bill Gates who had a bright idea for software, the California electronics tinkerers who figured out how to make the machine user-friendly, and the two friends who assembled the first real desktop machine in a cluttered garage. They put an apple logo on their first machine, a symbol very few understood: When Alan Turing, the real father of the computer, decided to end his tortured life, he prepared some cyanide in his own laboratory, injected it into an apple, then took a fatal bite. What message Turing was trying to convey by this method of death, fraught with biblical overtones, can only be guessed.

• • •

More than any other scientific or technical development, it was the computer that formed the cutting edge of the superpower of superpowers, the United States. American might had come about because of a conviction that science was the key to advanced technology and that, in turn, was the key to military dominance. Put simply, from the moment the smoke began to clear from Pearl Harbor, the United States decided it would never again be a second-rate military power. Only a lavish investment in scientific research, it was decided, could produce the advanced technology that in turn produced weapons superior in sophistication and power to make the United States invulnerable to any threat. This was not a new idea, of course; beginning in a time before the flowering of ancient Greece, science has had a close relationship to military power. By the twentieth century, very few doubted that military power was grounded in the applications of advanced scientific knowledge to military technology.

But that required money, a lot of it. The United States had the money to underwrite the development of the nuclear reactor, the microchip, the transistor, and solid-state electronics, the essential building blocks of the scientific-technical revolution that transformed war. The Soviet Union did not have the money, and that, fundamentally, is why it could not sustain the scientific race with the United States. Through supreme effort, the Soviets could achieve occasional spectaculars, such as launching the first man into space, but they were hollow; showy spectaculars hid what was at root a poor scientific base, starved for money. While the Soviet Politburo was congratulating itself on triumphs of Soviet scientists, those very same scientists were finding that they didn't have enough money to buy such simple laboratory tools as toothpicks, or tissue to clean their instruments.

The Soviet Union fell into the dustbin of history, the penalty exacted against ambitious powers whose reach exceeds their grasp. As history teaches, however, there are other nations seeking to fill that gap in the ranks of great powers. And since history tends to repeat itself, two ambitious powers whose ambitions echo far back in their history have sought to recapture their ancient glory, trying to relearn an equally ancient lesson about science harnessed to military power.

●   ●   ●

The nuclear explosion that lit up the Gobi Desert in 1964, the test of China's first nuclear weapon, signaled to the world that China had emerged from nearly two centuries of backwardness and domination by foreign powers. Its meaning was unmistakable: Chinese science had been reborn and directed toward reestablishing the might and the power of ancient China.

However, appearances were largely deceiving. In fact, the Chinese used mostly borrowed Soviet and other foreign technology and scientific expertise to bring forth a spectacular event calculated to vault China into the rank of superpowers. The effect did not quite come off, for Chinese science was clutched in the deadening hand of Mao Zedong, a scientific illiterate whose "cultural revolution" propounded a theory of peasant genius triumphant over such class enemies as university professors and anyone with a higher learning degree. Mao, the man who had dispatched Chinese troops without air cover or armor against American firepower in Korea on the theory that his peasant soldiers could defeat modern military technology by sheer force of will (a monumental act of stupidity that cost the Chinese more than a million casualties), wanted to re-create the glory of ancient China. But without science, he had no hope of achieving it—as his successors realized. In 1978, three years after Mao's death, they announced the "Four Modernizations," which amounted to nothing less than a vast bootstrap program to bring Chinese science on a par with the United States. Huge sums were sunk into basic scientific research, more than a hundred professional science societies were formed, and tens of thousands of students were sent abroad, at government expense, to study science at the leading Western universities.

This drive, intended to bring about a renaissance of Chinese science virtually overnight, is linked to China's other major push, military modernization. For more than twenty years, the Chinese have been engaged in an unprecedented military buildup to construct a military machine for purposes not quite made clear, with all the components of modern military power—ICBMs, satellites, missile-firing submarines, thermonuclear weapons, high-performance jet warplanes. All these components have been turned out by a growing scientific-technical structure designed and controlled by the state.

A similar process took place in Iraq as it sought to recapture the glory of the ancient Hittite and Babylonian superpowers whose ruins can still be seen there. When Saddam Hussein gained total control over the country in the 1970s, he moved to create a modern Hittite empire using the same playbook that had made the Hittites so formidable: ruthless control to create a garrison state armed with advanced science and technology. Huge oil revenues allowed Hussein to create, within only a few years, a scientific-technical complex designed solely for military power, including work on nuclear weapons and chemical and biological weapons. The process, aided by an infusion of Soviet military technology and expertise, convinced Hussein he had at hand a new Hittite military machine that would allow him to create a new Middle Eastern empire, just as the Hittites had done.

But like Hitler, Stalin, and any number of other tyrants, Hussein was a scientific illiterate who paid very little attention to what was going on outside his own universe. While he was infatuated with all the technology that led him to believe he was invincible, he failed to realize that in modern war, he who has the better technology wins. And he failed to realize that the United States had the better technology, which it used, to devastating effect, to overwhelm Hussein's inferior technology. It was a lesson Hussein learned the hard way in the first hour of the Gulf War in 1991, when that superior American military technology wiped out what Iraq thought was a state-of-the-art air defense system, leaving the country hopelessly vulnerable to air attack. Underscoring the rapid pace and tremendous violence of modern war, it took only sixty minutes for superior American science and technology to obliterate a system that had taken fifteen years and more than $9 billion to build.

The unforgettable video images of that war—missile warheads with television guidance recording their destructive power as they entered ventilation shafts of targeted buildings, laser-guided bombs hitting targets with pinpoint precision, Iraqi tanks picked off in the darkness by helicopter gunners using night vision devices, "stealth" planes invisible to radar—represent the culmination of a process that began at the dawn of civilization and that has been accelerating ever since. It is a process that combines two of the human species' most unique characteristics: an instinct to dominate fellow members of the species, along with an obsession to understand the natural world and to dominate that, too. The urge to dominate became war, and that drive to understand the natural world (and reshape it to mankind's ends) became science. Ever since, they have been walking hand in hand through human existence, with consequences that now threaten the very future of the planet.

The ancient Greeks seemed to have sensed the approaching danger, the moment when the search for knowledge would become the search for death. Their myth of Prometheus resonates to this day because it encapsulates, in Prometheus's punishment for stealing the secret of fire, the uneasiness we feel about the relentless search in nature's secrets for better ways to kill each other—and in ever greater numbers. Judging by the current scientific research into future weapons, wars to come will tap into the deepest secrets of nature, the Promethean fire that so concerned the ancient Greeks— microbes, dispersed by bombs, to literally eat stores of oil and gasoline; bombs and shells to suck up all the oxygen in a given area, asphyxiating all living things; powerful X-ray lasers capable of destroying any object with a "death ray"; streams of charged particle beams fired by guns at speeds up to 3,000 yards a second; nanotechnological weapons the size of molecules to infect and then destroy human beings.

If history is any guide, it is a virtual certainty that these weapons—or some form of them—will come to pass because history also teaches us that war is inevitable. And then what? Nobody quite knows, although the twentieth century's history, when the industrial revolution became a technical revolution, offers some guidance. Most importantly, technology must be controlled. It was informed political action that cured the worst ills of the industrial revolution, it was informed political action that ended the unrestrained use of agricultural chemicals, and it was informed political action that ended atmospheric testing of nuclear weapons. And unless the use of science in war is restrained and controlled in some way, even the smallest war threatens to spin out of control, with an impact far beyond whatever local border dispute or other problem set it off (e.g., consider the worldwide consequences of a localized war between India and Pakistan, two nations now armed with nuclear weapons).

Should such a disaster come to pass, it would fulfill the dark vision once propounded by the physicist J. D. Bernal, who predicted a time when science would so dominate the conduct of war that the complexity of the science necessary to fight it would lead to the scientists actually running governments. Given the growing power of science's role in war, that vision does not seem entirely out of the realm of possibility, although it should be kept in mind that prediction, as Niels Bohr once noted, "is very difficult, especially about the future." So we don't know if there will be an increase in the estimated five hundred thousand scientists and technicians worldwide who now devote their time exclusively to devising more deadly ways of waging war, or, to cite a parallel statistic, if there will be an increase in numbers beyond the several thousand scientists who have signed pledges never to do any scientific work on space-based missile defenses.

And above all, we don't know if Einstein was right when he said he couldn't predict with what weapons World War III would be fought, but he was certain of the weapons that would be used in World War IV: stones and clubs.

Of Microbes and Thunderbolts

If a little knowledge is dangerous, where is the
man who has so much as to be out of danger?

—T. H. HUXLEY

The first man in space had no astronautical or flight training of any kind; indeed, he had never even been on an airplane in his life. In the spring of 1942, the man known to history only as "L," reportedly a Jewish delicatessen clerk, was shoved into a decompression chamber at the Dachau concentration camp in Germany. For the remaining few minutes of his life, he was subjected to agonies beyond human comprehension as the air pressure inside the chamber was lowered to a level approaching outer space. When the mercy of death finally ended his torment, his body was dragged out of the chamber and thrown into a crematorium. His ashes were shoveled into a ditch, there to join the ashes of thousands of other anonymous victims.

To the dozen men in white laboratory coats who impassively watched the horrible death of "L" (and recorded it with a movie camera), he was just so much flotsam to be discarded in the name of what they called "scientific research." Their human subject was an *Untermensch* (subhuman), selected precisely because as a concentration camp inmate, he had no status beyond that of a white laboratory mouse. He was among several hundred Dachau inmates herded into a secluded building in the camp that spring and sacrificed to the exigencies of war.

Their fate had been decided several months before, when the scientists of the *Deutsche Versuchsonstalt für Luftfahrt* (German Experimental Institute for Aviation) approached Heinrich Himmler with a problem. Pilots of new German aircraft with more powerful engines and able to operate at higher altitudes were encountering difficulties: As they approached 50,000 feet they suffered various physical problems and had trouble functioning, even with a supply of oxygen pumped through face masks. These problems would have to be solved, especially considering that designs for planes capable of operating up to 100,000 feet in altitude were already on the drawing boards, a high-speed jet fighter was already in flight testing, and there was talk of putting men in the noses of the new long-range rockets under development. The urgent tasks were to conduct research on the effects of high-altitude flight on humans and to devise some kind of system—possibly a pressurized crew compartment—that would allow pilots at high altitudes to function efficiently. A crash research program had begun in 1941, but tests using monkeys hadn't worked. The only solution, the men of the *Versuchsonstalt* decided, was human subjects, but unsurprisingly, they had not been able to recruit any volunteers. That led them to Himmler's office door: Could the *Reichsführer SS* provide the necessary "volunteers"? Himmler certainly could; he had millions of subhumans behind barbed wire, and since he intended to kill them all anyway, he was perfectly willing to lend out some of his victims for medical experiments. Shortly thereafter, a "scientific research center" was constructed at Dachau, and the first "selectees" were dragged there to undergo their final agonies.

The men responsible for those sufferings were not monsters or sadists. They were in fact the cream of German medical science, inheritors of a distinguished tradition that had alleviated much human suffering. Now, however, they had become murderers. And there were plenty of compliant coconspirators among their distinguished colleagues: Nearly two hundred of Germany's most prominent doctors, surgeons, pathologists, and microbiologists sat mute several months later when, at a scientific conference in Nuremberg convened by the *Versuchsonstalt* to review progress of research into the problems of high-altitude flight, the results of the Dachau experiment were outlined in dry scientific language. They were shown movie film of "L," clad in his concentration camp striped pajamas, writhing and screaming in agony as he died in the Dachau chamber; an inset showed an air pressure gauge, so that viewers could see at what point his body stopped functioning. Another topic on the conference's agenda was an equally horrifying series of experiments that were conducted at the same time in Dachau. These experiments, designed to investigate the problem of hypothermia suffered by Luftwaffe pilots forced to bail out into frigid ocean waters, featured dozens of Dachau inmates who were thrown naked into tanks of ice-cold water and forced to stay there until their body temperatures dropped to near-death. At that point, the inmates were removed from the tanks and put into beds, where naked Soviet female prisoners of war were forced to embrace them in an attempt to determine if intimate human contact could reverse the effects of hypothermia. Other inmates were hosed down and forced to stand naked outside in frigid winter temperatures in an effort to determine how long it took human beings to freeze to death.

Eighty Dachau prisoners died and more than a hundred were crippled for life by the experiments, but there were no cries of outrage when the conference attendees heard of these atrocities. None of the distinguished medical scientists in attendance walked out in protest, or demanded that such experiments be ended, or registered even the slightest qualm of conscience. Perhaps they took their cue from the man presiding over the conference, Germany's preeminent physiologist (and head of the *Versuchsonstalt*), Dr. Hubertus Strughold. With the cool scientific detachment of a researcher reciting accounts of experiments involving dissections of dead frogs, Strughold conducted a discussion on what German science had learned from the Dachau experiments, and laid out further "lines of research" (presumably to be conducted at the expense of other Dachau inmates).

Tragically, Strughold and his fellow paragons of the German medical science establishment were not alone in this abyss. Hundreds of their colleagues of the world's most distinguished scientific elite were simultaneously busy perverting the principles and ideals of their chosen disciplines—psychiatrists and medical doctors who murdered the mentally ill because the state claimed that such people were "unfit to live"; chemists who developed Zyklon-B poison gas—"insecticide for people," as they called it—because the state wanted to murder millions of *Untermensch*; anthropologists who propounded crackpot theories of "Aryan racial superiority" for Himmler's SS; epidemiologists who deliberately infected thousands of concentration camp inmates with deadly diseases in tests to develop toxins for Germany's biological weapons program; and, most horrifying of all, experiments on children in concentration camps to test various bizarre medical theories (such as whether blue eyes could be implanted from one child to another).

How could it have happened? How could so glittering an array of talent and genius become so thoroughly perverted in service to a criminal state?

The answer has everything to do with the final, complete mating of science and war in the twentieth century. What happened in that century ended at last the traditional and persistent claim of science that it is strictly an objective reality independent of laws, customs, authorities, and loyalties, and that it is divorced from the exigencies of the state. The great power struggles of the century, arising largely from the creation of ambitious nationalist states, shattered forever the idea of science as a pillar of independent thought and science as an independent entity. In a time when belief in and adherence to the social and political institutions of the state became the new religion, science did not remain independent. Science became indispensable to the state; the modern world's Merlins were the wizards who produced the magic machines that won wars. To a large extent, the arrangement was Faustian: scientists were pampered, sheltered, and rewarded as long as they were unswervingly loyal to the state. They were to devise the science and technology to maintain the power and existence of the state, with the understanding that their science belonged exclusively to the state and that they were to function strictly and unquestioningly as state functionaries. Very few scientists resisted this process, most notably Albert Einstein (who refused to allow his genius to be used for the development of weapons) and Guillermo Marconi (who refused blandishments from his native Italy to use his breakthroughs in radio technology in the service of Mussolini's regime).

It required only the creation of a criminal state, such as Nazi Germany, to demonstrate how dangerous this mating of science and state could be. But it would be too easy to dismiss what happened there as an aberration in the history of science, the tragic consequence that results when a totalitarian state bends science to evil ends. The sad fact is that Nazi Germany's perversion of science was not an isolated event; witness what happened elsewhere.

• • •

At about the same time that German medical science was murdering Dachau inmates, some 6,000 miles to the east, an even deeper descent into evil was taking place at a Japanese prisoner of war camp near the dusty Manchurian city of Harbin. There, men assigned to Unit 731 of the Imperial Japanese Army Water Supply and Prophylaxis Administration one night injected three thousand bread rolls with typhoid germs they had grown in a laboratory. The next morning, the rolls were fed to unsuspecting Chinese prisoners. Within days, thousands of them were writhing in agony as an epidemic sucked the life from their bodies. There was no attempt to mitigate their suffering or cure the disease that was destroying them; the white-coated Japanese doctors in surgical masks who moved among the dying prisoners instead meticulously noted how rapidly the disease took hold and which physical types succumbed the quickest.

As in Germany, the men in white laboratory coats who murdered thousands of prisoners were not monsters or sadists. They were, in fact, among Japan's most prominent bacteriologists and microbiologists, men who had originally enlisted in a scientific discipline dedicated to the creation of immunity against fatal diseases and lessening the virulence and toxic nature of microbes. But now they were involved in Japan's darkest secret, an attempt to use those microbes for biological weapons capable of wiping out millions of human beings. As scientists, they were very much in the mold of the man who had recruited them for this ultimate union of science and death, Dr. Ishii Shiro, Japan's most prominent bacteriologist.

Shiro was the product of a proud scientific tradition, born during Japan's war with China in 1894. Appalled to learn that 80 percent of Japan's casualties from that war were caused by disease, the Japanese government ordered a major scientific effort to reduce disease on the battlefield. Japan's best microbiologists, medical researchers, and epidemiologists were enlisted for the effort, which resulted in a dramatic reduction in battlefield diseases. In the process, the Japanese devised a number of innovations later adapted by all civilized nations, including field hospitals, medical personnel assigned to front-line combat units, and extensive sanitation procedures. Japan's efforts in this area led to major advances in bacteriology and combating epidemics, and by the early twentieth century, Japanese epidemiology was ranked the world's best.

Shiro, who had made his mark in world-class bacteriological research, was, like Strughold in Germany, enlisted by his nation's military to utilize his scientific talents for war. Initially those talents were concerned with improving Japan's already stellar reputation for reducing battlefield disease, but by 1930, the Japanese military had become preoccupied with a very dangerous idea: deadly microbes as weapons of mass destruction, the supreme weapon for a small nation up against much larger rivals. Shiro, a fanatical nationalist, agreed to head a top-priority research effort, to be conducted in deep secrecy. In 1932, when Japan invaded Manchuria, Shiro recruited Japan's top bacteriologists and microbiologists for the research effort, concealed behind the deliberately vague nomenclature of a military sanitation unit. At a secret complex, Shiro and his fellow medical scientists performed horrible experiments on Chinese prisoners, including vivisection, to determine if they could manufacture sufficient numbers of toxins and microbes, along with delivery systems, for an efficient weapon of mass destruction. (Later, similar experiments were performed on Soviet POWs and American soldiers who had been captured in the Philippines.)

By 1940, Shiro had developed the world's first biological weapon of mass destruction. It was utilized in attacks on Chinese villages; one village of four thousand people was wiped out by bombs containing rats with bubonic plague–infected fleas. In 1945, an attempt to wipe out U.S. forces invading Okinawa failed when a ship carrying pathogens to the Japanese defenders was sunk by an American submarine en route. An even more ambitious plan that year—to drop pathogens on Los Angeles, using seaplanes launched from a giant submarine—was vetoed by the chief of the Japanese General Staff, who noted, "Japan will earn the derision of the world." To no avail, Shiro argued that since the Americans were using napalm, a "hideous weapon," to incinerate Japanese defenders in deep fortifications, Japan was justified in using a "weapon of terror" to kill millions of Americans in Los Angeles.

Ironically, Shiro might have won the argument had he known that at that very moment, at a secret site in New Mexico, the Americans were developing their own weapon of mass destruction. He might further have argued that in an age of total war—in which entire populations were mobilized—the line between civilian and military had become increasingly blurred, to the extent that the civilian worker who assembled a machine-gun barrel was regarded as legitimate a target as a front-line soldier. And war had taken an even more fateful step: slaughtering mass numbers of civilians for the crime of being citizens of an enemy nation. There was no military justification for the German bombing of London and the Allied destruction of Dresden (where more civilians died than were killed at Hiroshima) save that of killing as many civilians as possible in an attempt to terrorize and destroy the morale of an enemy nation's people.

In that atmosphere of unrestricted slaughter, it could be argued that there was only a qualitative difference between using poison gas or incendiary bombs to kill people. Both were terror weapons; one choked its victims to death, the other created a firestorm that turned air raid shelters into tombs. What, then, was the difference between the scientists who devised poison gas to kill mass populations and the scientists who built "city buster" bombs to accomplish the same end? The scientists had become simply soldiers for a war in which they had no choice but to fight, however unethically. In an age of total war, these scientists faced the ultimate ethical dilemma: Either use their scientific talents against ruthless enemies, or stand aside and let their native lands suffer total destruction. True, they could have followed the example of Marconi, but his act of refusal did not deter Mussolini in the slightest from invading Ethiopia, or joining Hitler's war of aggression.

Nevertheless, it was not an argument accepted by the victorious Allies after World War II. In a series of postwar war crimes trials, the "Nuremberg Doctrine" was established: No one can cite "I was just following orders" or the exigencies of the state, however threatened it might be, to commit crimes against humanity. For science, the doctrine mandated that no scientist can violate scientific ethics—such as experimenting on unwilling human subjects—in the name of the state, no matter the circumstances. Fundamentally, Nuremberg proscribed limits on the relationship between science and the state, particularly in time of war. To be sure, it accepted the role of science in modern weapons of mass destruction; no American nuclear scientist was charged with war crimes for devising a weapon that killed more than a hundred thousand unarmed civilians in Hiroshima.

But the ink was hardly dry on the Nuremberg Doctrine when its propounders demonstrated the very same moral myopia that had led German and Japanese scientists into an abyss. It was a myopia that would permanently stain the science establishments of the victorious democracies for many years to come—and would assume its most virulent form in the cases of two major war criminals named Dr. Hubertus Strughold and Dr. Ishii Shiro.

. . .

After the end of the war, Strughold returned to academia, becoming professor of physiology and head of the Physiological Institute at the University of Heidelberg, despite his inclusion on an Allied list of Nazi war criminals to be arrested and tried. Several dozen German scientists involved in the Dachau experiments and other atrocities had been rounded up and were on trial before the International Military Tribunal, but Strughold, their boss, remained mysteriously free. What only a few people knew was that he was under the protection of a dark American secret.

The secret was code-named Operation Paperclip, an intelligence operation that grew out of a colossal American scientific mistake. The government had exempted scientists from the military draft but did not exempt science students. After war's end, the United States realized its great mistake: From 1942 to 1945, some ten thousand doctorates in science had been lost. Now locked in a new struggle with the Soviet Union, something very close to a full-fledged panic broke out at the highest levels of the American military and political establishments. Not only was there a shortage of scientists—modern war's indispensable wizards—but also, the Americans discovered, to their shock, that their science, with military implications, despite its remarkable record in World War II, was seriously behind. American scientific intelligence teams that fanned out all over Germany in the wake of the invading armies discovered that German science (except nuclear science) was decades ahead of American science. The German superiority extended across the entire scientific spectrum, most notably rocket science: In December, 1944, when German V-2 rockets were smashing into London from launching sites hundreds of miles away, the Americans were testing their most advanced rocket weapon, a small surface-to-surface missile called the Private A, which achieved a range of 11 miles. The scientific intelligence teams also discovered that the Germans had tested a submarine-launched version of the V-2 to be fired into New York City, and had developed antiaircraft missiles, advanced swept-wing jet aircraft, intercontinental bombers, and an advanced arsenal of chemical and biological weapons (which Hitler decided not to use for fear of massive Allied retaliation). It was, the Americans realized, a near-run thing; only German disorganization and lack of resources had prevented Hitler's "wonder weapons" from winning the war.

There was an even greater shock in Japan, where American teams discovered Shiro's biological weapons program, an arsenal that assumed even more frightening proportions when contrasted with America's elementary chemical and biological warfare (CBW) arsenal—which consisted, mainly, of large stockpiles of World War I–era mustard and nerve gases. Another close-run thing; if the Japanese had used their biological weapons against American forces—or had been able to develop their nascent nuclear weapons project—then the outcome of the war in the Pacific might have been very different.

While most Americans assumed that their science and technology—exemplified by such wonders as the atomic bomb and the B-29 bomber—were vastly superior to that of their defeated enemies, the scientific intelligence teams knew better. Their alarming reports led to Operation Paperclip, which sought to achieve two objectives: one, buy time by using captured scientific assets for a rapid buildup of America's science while its educational system made up the wartime deficit in science graduates; and two, deny those captured assets to the Soviet Union. The result was the greatest transfer of scientific talent in world history: Several hundred German scientists (along with Ishii Shiro and the top scientists of Unit 731) were secretly whisked away under American control and enlisted to build a new American scientific-technical military monolith.

But the operation required an acute moral myopia by the American government. For one thing, American law strictly prohibited any member of the Nazi Party, those involved in "crimes against humanity," or anyone affiliated with "criminal organizations" (defined as the Nazi Party, SS, Gestapo, and other Nazi organizations) from entering the United States. For another, the United States was a signatory to agreements among the Allies pledging the victorious nations to bring all war criminals to justice, the purpose behind the International Military Tribunal. These legal barriers represented a real problem, since it was discovered that the majority of the German and Japanese scientists sought for Operation Paperclip were either "members of criminal organizations" or outright war criminals subject to prosecution.

The problem was solved with a criminal act: forgery. The personnel records of those scientists sought by Operation Paperclip were marked with a paper clip (hence the operation's code name) and removed from the files of pending war crimes prosecution procedures and de-Nazification trials. Then documents were prepared reporting that "extensive investigations" had determined that a particular Operation Paperclip recruit had never been affiliated with the Nazis in any way and had never participated in war crimes. Of course, no such "investigations" ever took place. The Americans involved in the operation were perfectly aware of the odious background of most of their scientist-recruits, but expediency prevailed: If a scientist was deemed important to U.S. national security, he was enlisted, no matter how many murders he had committed.

The recruitment process went smoothly, and no wonder: The Americans hardly needed to point out that the scientists recruited really had little choice. Any scientist who decided he did not want to work for a new master faced the alternative option of facing a war crimes investigation. No surprise, then, that several hundred scientists, most of them German, quickly agreed to enroll in Operation Paperclip, with its incentives of immunity from war crimes, good salaries, secure government jobs, and American citizenship. Among them was Dr. Hubertus Strughold.

In 1947, Strughold was brought to the United States under Operation Paperclip. He was regarded by the U.S. Air Force as a prize catch, since he was the world's leading expert on the physiological effects of high-altitude flight. Moreover, he was also the world's leading expert on the then-adolescent science of the physiology of spaceflight. There were few air force officers unaware of how Strughold had gained this expertise, but in an atmosphere of expediency, that was not a matter considered worthy of discussion. Judged strictly in scientific terms, Strughold would make invaluable contributions to the American military and spaceflight programs. He was christened as "the father of space medicine," and his innovations include pressurized flight suits for high-altitude flights, artificial atmosphere systems for space capsules, and the discovery that zero gravity could be created to train astronauts (achieved by putting a plane into a power dive, then pulling out, pointing the nose up, and throttling down the engine to minimum torque, creating thirty-two seconds of zero-G force). As a result, he was lavished with honors, including directorship of the U.S. Air Force School of Aviation Medicine. More ominously, a grateful military establishment protected him from a dangerous threat in 1964, when his name popped up on a U.S. Immigration Service list of Nazi war criminals living in the United States—all of whom were subject to deportation because their entry into this country was illegal. Following high-level pressure—including, reportedly, several high-level phone calls by President Lyndon Johnson—the Immigration Service quietly removed Strughold's name from the list.

A similar moral myopia was required for some of the most famous Operation Paperclip scientists, the "rocket team" of Germans who had built Nazi Germany's missiles and who would go on to create an American version that ultimately landed man on the moon. The problem was that the German V-2 program had utilized slave labor, most notoriously in 1943, when production was shifted from an open-air testing and production site on the Baltic coast (wrecked by British bombing) to an underground factory inside a mountain in southern Germany. Outside the underground site, the Germans built a concentration camp called Dora, into which they brought sixty thousand inmates from other concentration camps as slave laborers. Under SS whips and cudgels, they hacked out huge production rooms by hand. More than half of them were worked or beaten to death; American troops who liberated them in 1945 wept at the sight of the skeletal survivors and the huge piles of broken bodies awaiting the ovens of a crematorium.

The man who asked Himmler for the slave laborers was the head of production for the V-2 program, Dr. Arthur Rudolph, a brilliant physicist who had designed the V-2 rocket engines. According to Dora survivors, he had also ordered public hangings of slave laborers accused of sabotage or refusing to work. Rudolph's odious record was well known to the Americans, but as the world's leading expert on rocket engines, he was a prime recruit for Operation Paper-clip. His record as a Nazi Party member and war criminal was ex-punged, and Rudolph joined his more famous working partner, Dr. Wernher von Braun, in the exodus of German science to the New World. Judged strictly in military terms, Rudolph was indispensable to both the American military rocket and space programs; his crowning achievement was the huge Saturn rocket of 7.5-million-pound thrust that launched the Apollo spacecraft to the moon.

If the great American triumph of landing on the moon has at least a partial heritage in a Nazi concentration camp, the much less-known U.S. C.B.W. program is grounded completely in evil, for it was almost entirely created by two war criminals, Dr. Ishii Shiro, head of Japan's Unit 731, and Dr. Walter Schreiber, the microbiolo-gist who created Nazi Germany's biological warfare arsenal, using toxic agents developed as a result of testing on concentration camp inmates. Shiro, obviously, was a prime candidate for prosecution at the Allied war crimes tribunal in Japan after the war. Indeed, the American prosecutor, in one of the early sessions of the trial of major Japanese war criminals, introduced evidence that during the Japanese occupation of Nanking in 1939, Chinese prisoners were subjected to experiments by Unit 731 personnel. But he then added, "We do not at this time anticipate introducing additional evidence on that subject."

This startling statement infuriated Soviet members of the tribu-nal, who were determined to place Dr. Shiro in a hangman's noose. When Soviet troops invaded Manchuria in 1945, they came across Unit 731's main laboratory, along with evidence of the unit's exper-iments on human beings (including hundreds of Soviet prisoners). Shiro and several other top leaders had managed to flee to Japan ahead of the rapid USSR advance, but several hundred personnel of Unit 731 had fallen into Soviet hands—along with an even more in-criminating stack of documents Shiro and his henchmen had no time to destroy.

The Soviets quickly discovered why the Americans had decided not to prosecute Shiro: He had made the ultimate plea bargain, agree-ing to build a biological warfare capability for the United States in exchange for immunity from war crimes. As a protest, the Soviets convened their own war crimes trials in Manchuria, which resulted in the execution of dozens of Unit 731 members. However morally superior the Soviets might have felt about this course of events, it soon ended as Cold War expediency took hold. Moscow was concerned about what was perceived as an American crash program to develop the world's mightiest CBW arsenal, one for which the Soviet Union had no defense. Not only was Shiro busily at work for the Americans, but also Dr. Walter Schreiber, the German microbiologist who had created Hitler's CBW arsenal, had turned up missing, and the Americans did not seem especially eager to help track down one of the top wanted war criminals. The Soviets were especially eager to bring Schreiber before the war crimes tribunal, since many Soviet prisoners of war had been sacrificed in his experiments on live human subjects. Schreiber wasn't missing; he was under American protection, having rapidly (and understandably) snatched up the American offer: help the development of American biological weapons or face war crimes charges. (By 1952, Schreiber finished his work with the Americans, who then faced the awkward problem of what to do with him. Too notorious to be admitted to the United States, Schreiber ultimately was hidden away: The Americans arranged his immigration to Paraguay, whose military dictatorship was already protecting a number of notorious war criminals, most infamously Dr. Joseph Mengele.)

Panicked over what they perceived as an ultimate American weapon of mass destruction, Moscow soon lost interest in any moral questions on the matter of biological weapons. Armed with the huge stash of research documents seized in Manchuria, the Soviets set about creating the largest and most sophisticated biological weapons capability in history. In an effort that extended to the fall of the Soviet Union itself in 1991, the Soviets siphoned off the cream of their medical establishment—the best biochemists, epidemiologists, medical researchers, and biologists—and set them to work in a number of vast complexes where toxins were developed and mass-produced. The effort set back Soviet medical science decades, as the lion's share of resources was devoted to work on biological weapons. As a result, Soviet hospitals were somewhere around the lowest Third World standards, and even such basic medical commodities as antibiotics and syringes were in critically short supply. Worse, the frantic effort to build what the Soviet military called its "biowar" arm caused a lowering of safety standards. As a result, there were several accidents, most notoriously the accidental release of anthrax toxins at a biological warfare research center in the city of Sverdlovsk. In a government cover-up story nobody believed, the resulting death of hundreds of civilians was blamed on a cow dumped into a well, thus contaminating the water supply.

The Soviets were equally panicked by Operation Paperclip. Determined to round up German rocket scientists and aeronautical engineers, along with their wizardry, Soviet scientific intelligence teams rushed into Germany just behind the troops who liberated Berlin, but to their fury discovered that the Americans were one step ahead of them. (At one aeronautical research laboratory in East Germany, the Soviets arrived to find that every employee had been taken westward by their American rivals, who had thoughtfully swept the laboratory floor clean for the new Soviet landlords.) Enraged that the Americans had stolen a march on him, Stalin ordered an all-out scientific effort to match the German World War II successes, then to match and exceed whatever American successes would arise from their recruitment of the German scientists.

To do so required solution of an awkward problem.

The problem was that the man who had ordered a crash program to develop Soviet rockets was also the same man who, in one of his more pronounced fits of paranoia, had thrown most of his rocket scientists into the Gulag some years before on suspicion that they were "anti-Soviet saboteurs" (meaning they had demonstrated too independent a turn of mind). Most prominent among them was Sergei Korolev, a mathematician and physicist who as a young graduate student began experimenting with launching rockets of his own design in his spare time. Like Wernher von Braun, he and other fellow rocket enthusiasts dreamed of the day when large rockets would carry man to the outer planets and when artificial satellites would circle the earth. And as in von Braun's case, these youthful dreams were soon preempted by the state, which decreed that Korolev's talents were better suited working in an airplane design bureau on future jet aircraft. In 1937, Korolev one night was taken to a NKVD prison, where he was accused of being "anti-Soviet," beaten to a pulp, then sentenced to twenty years in a Siberian labor camp. In 1945 he was suddenly "rehabilitated" when Stalin ordered Soviet science to match the German V-2 rockets. Korolev was put in charge of a Soviet scientific team that investigated the V-2, and when he announced he could match, and then better, the V-2 design, he was put in charge of the Soviet military rocket development program. Within a decade he had developed the R-7, a multistage rocket with a 6,400-kilometer range, Russia's first intercontinental ballistic missile. In 1957 he designed Sputnik and the rocket that carried it into space, a remarkable scientific feat considering the Soviet Union's persistently inferior technology. (But that inferiority prevented Korolev's greatest dream, landing Soviet cosmonauts on the moon. The Soviets could not develop the miniaturized, low-power electronic and computer systems that made American rockets capable of long-term spaceflight.)

Given Korolev's history, it remains a mystery how he could have subsequently devoted his life and talents to developing weapons for a government that had so badly mistreated him. And there is an even greater mystery: How could a man who knew firsthand the horrors of the Soviet system consent to serve a criminal state? How could he have overlooked the dark side of the Soviet rocket program, the thousands of slave laborers who were worked to death, virtually outside his office window, building the vast cosmodromes and other facilities needed for rocket research and launchings? How could he have remained silent when so many of his friends in the Soviet scientific community were dragged off to prison on assorted trumped-up charges—including his close friend the geneticist Andrei Vavilov, who was arrested and hideously tortured until he confessed to being an "English spy," then sent to a Siberian labor camp, where he died of malnutrition and exposure? (Vavilov's "crime," as it turned out, was being elected to the presidency of the 1939 International Congress of Genetics in Edinburgh, an honor that somehow convinced Stalin that Vavilov was spilling the secrets of Soviet genetics to British intelligence.)

Korolev took the answer to the grave with him when he died in 1966, but it almost certainly had to do with political expediency, the single greatest cause of evil in this century and the single greatest reason for the perversion of science. Despite all that he underwent at the hands of the state, Korolev remained a fervid Russian patriot who believed that when it came to the subject of the survival of the state, anything was justified. He believed, as did many of his countrymen, that Russia (as distinct from the Soviet state) was in grave danger from its enemies, and that science provided the means to achieve invulnerability to those threats. The young graduate student who dreamed of rockets flying men to explore the solar system became the hard-hearted scientific bureaucrat who allowed slave labor to help build a huge force of intercontinental missiles with thermonuclear warheads. The young scientific idealist who hesitated sending white mice aloft in one of his early rockets for fear of having the mice suffer, years later would react impassively to the heartrending radio messages of a Soviet cosmonaut screaming in agony as his malfunctioning capsule was incinerating on reentry, then order his workers to "clean up the mess" and get another cosmonaut ready for the next flight.

As in the case of Nazi Germany and the militarist Japanese Empire, it is too easy to dismiss the complex case of Korolev—and other Soviet scientists—as an aberration in the history of science, still another example of what happens when science is taken over and perverted by a totalitarian state. In this line of reasoning, what happened in Germany, Japan, and the Soviet Union represents extreme and isolated examples of the twentieth-century phenomenon of the ultimate politicization of science. Yet the fact is that the very same process took place in the one place where it shouldn't have happened, in the world's leading liberal democracy, the United States.

• • •

Even in retrospect, it seems hard to believe: At virtually the same moment when American prosecutors in Nuremberg were charging Germany's medical science establishment with crimes against humanity for performing tests on unwilling subjects in concentration camps (seven German scientists were hanged after being found guilty), American science was doing precisely the same thing. For nearly forty years, the American scientific establishment conducted:

- tests involving deliberate exposures of radiation to nearly twenty thousand unwitting patients to determine the physiological effects of radiation, including a group of mentally retarded children fed radioactive iron;
- secret releases of clouds of radioactive iodine in the atmosphere over Washington State to judge scientists' ability to track radioactive plumes;

- the feeding of radioactive cereal to unsuspecting youngsters at institutions for troubled adolescents to determine how radiation was deposited in the human body;
- tests that involved giving psychochemicals—including LSD, BZ, and other hallucinogens—to unsuspecting recipients to determine if such drugs would allow control of human beings;
- tests to determine the practicality of various biological warfare weapons by secretly releasing bacteria and other germs in various populated areas, including lightbulbs filled with germs in the New York City subway system.

These incidents represent just a few of the many thousands of similar tests, all conducted under official government sanction and supported by government funds. And they were all conducted under the mantle of expediency: The United States faced the gravest threat in its history from the Soviet Union, so virtually anything was justified in the name of national security. In a tragic irony, the very same justifications that German scientists used to justify their experiments on unwilling subjects—the nature of the threats jeopardizing the very existence of Germany as a nation and a culture—were advanced by American science. Thousands of American scientists were perfectly aware that what they were doing violated the laws of humanity and their own nation. The American nuclear physicists with advanced degrees from Princeton, MIT, and the California Institute of Technology knew that participating in a test that paraded thousands of American soldiers near a nuclear explosion while assuring them they would come to no real harm as a result was a crime; unless they had slept through their classes in basic nuclear physics, every one of those scientists was aware that nuclear radiation damages human bodies. But the physicists could cite the pressure of expediency; it would have been virtually impossible to find willing human subjects for tests of radiation's effect on combat soldiers, so the only solution they could think of was to find those willing subjects by lying to them. The necessity of determining whether soldiers could function on a future battlefield perceived to be dominated by so called "tactical" nuclear weapons, a circumstance that might determine the outcome of World War Three, represented a basic issue of survival that justified any means—

including the very tenets the United States sought to establish at Nuremberg.

The nuclear physicists were not alone in this moral and ethical swamp. The Cold War struggle between the United States and the Soviet Union, the greatest in history (short of a hot war), so dominated the societies of both superpowers that virtually the entire spectrum of science was sucked into the vortex. Science willingly went along; that's where the money, government support, honors, and fame were. The phenomenon reached into every corner of science—witness how the field of mathematics assumed a national security importance far beyond what any mathematician could have dreamed.

The reason had to do with the most important (and least known) aspect of national security, protecting government electronic communications and at the same time reading everybody else's. From the time, only some sixty years ago, when the process consisted of small groups, armed with paper and pencil, sweating over complex ciphers, it has grown into a vast monolith that continues to proliferate. As the electronic communications web becomes more and more pervasive, the effort to monitor and read it all has led to huge, sprawling complexes. The National Security Agency (NSA), for example, now has some thirty-eight thousand employees and 11 acres of supercomputers, including models that can perform 70 quadrillion calculations every 5 seconds.

What made such sprawling complexes possible is science—the mathematicians who devise advanced ciphers and methods to crack other ciphers, the computer scientists who design the supercomputers, the electronics scientists who design cipher-producing machines. For years, the NSA, armed with lavish government funds, has sought to recruit the cream of America's most brilliant mathematics graduates and has underwritten virtually every advanced computer development.

However impressive scientifically, these advances contain any number of implications that have not yet been resolved. For example, the NSA managed to gain access to the computer systems used by Iraqi dictator Saddam Hussein's military before the Persian Gulf War and insert a "trapdoor" in the computer programs. With that entry point, the NSA was able to read all Iraqi computer transmissions, including the enciphered ones, as the war broke out. It was the key reason why American warplanes were able to pinpoint Iraqi air defense systems with such accuracy. But if the government has the capability to insert "trapdoors" in computer programs—including those used to encrypt communications—what safeguards exist to ensure that the same capability isn't used to read what's on everybody's hard drive? There aren't any, nor are there any safeguards to prevent the NSA—which now can vacuum up virtually every electronic signal in the atmosphere, including e-mail—from reading the private communications of citizens who are not drug dealers, terrorists, or spies.

But there is another, even more serious, problem such sprawling scientific efforts have spawned: proliferation. For decades, the science establishments of major powers have sought to keep major advances to themselves—only to discover that the task is impossible. Failing to heed the lessons of history taught since the fourteenth century—that if there is a scientific advance with important military implications, it will inevitably proliferate everywhere—key advances have been wrapped in deep secrecy in the belief that universal scientific principles somehow can be monopolized. However, as the nuclear physicists learned the hard way, that isn't possible. They wanted to keep the secrets of nuclear weapons among a small group of powers, but were helpless to prevent their colleagues in India from figuring out how to extract weapons-grade plutonium isotope 239 from the spent fuel of civilian nuclear power reactors and build nuclear weapons. Somehow overlooked was the fact that Indian scientists read the same textbooks and scientific papers as their American, French, British, and Russian counterparts.

The Indian obsession to build a nuclear weapons capability, diverting vast resources better used to ease the nation's serious social problems, underscores how the modern scientific-technical revolution dominates political thinking. That revolution promises a whole range of ultimate weapons to guarantee national security—nuclear-tipped missiles to guarantee North Korea's existence and goal of reuniting Korea by force, biological weapons to guarantee Iraq's status as a strategic military power in the Middle East, nuclear weapons in Pakistan to guarantee a checkmate of Indian territorial ambitions, thermonuclear-armed intercontinental missiles to guarantee China's dominance of the Asian mainland, space-based weapons to guarantee America's invulnerability to missile attack.

An estimated half a million scientists, representing every scientific discipline, are now engaged in strictly war-related work. Only a small portion of them have turned their backs on this largesse of government grants and contracts, signing avowals that they will not work on *Star Wars* missile defense systems, or demanding that physicists refuse any further work on nuclear weapons. The theory is that if science divorces itself from working on weapons, then those weapons won't be built. But again, history is instructive: As long as science promises invulnerability, a wonder weapon, or the means to dominate, then science will be seduced or dragooned to devise even greater weapons.

Those advocating a divorce between science and war face the Herculean tasks of not only fighting history but also the sheer scale of the modern scientific-technical revolution—and the implications of even the most nonmilitary developments. Consider the U.S. Department of Energy's research project to examine the radiation damage on the genetic systems of the victims of Hiroshima and Nagasaki. It evolved into a huge, sprawling project to establish the sequence of the estimated three billion base pairs in human DNA. The Human Genome Project, as it became known, promises major advances in medical science, but it also has immense military implications. Already studies are under way to determine how to manipulate genes, creating the ne plus ultra of weapons. Millions of dollars in research grants are being spent to see if new biotechnical weapons can be developed, substances that literally will destroy the genes of millions of people with only a few whiffs from an aerosol can.

And there are even more momentous possibilities lurking in dozens of other new scientific advances—genetic engineering, robotics, and nanotechnology (manipulation of matter at the atomic level to build molecule-size "assemblies" to cure cancer, clean up the environment—and provide a weapon that literally promises to rearrange all matter). As science continues to probe ever more deeply into the secrets of the microcosm and the macrocosmos, the range of destructive possibilities increases exponentially. How that process can be controlled remains the most vexing question confronting science; the deeper science looks into nature, the more ominous the possibilities that exist to destroy it. As long as science is wedded to the state, those possibilities remain to threaten all human existence.

It is a long road that has led us to this point, and we cannot say there weren't any number of warning signs along the way. Among them was the moment America's first great scientist, Benjamin Franklin, arrived in Paris to take up his post as American minister in France after the American Revolution. Franklin was a political and scientific hero to the French people, who turned out in droves to applaud the man they regarded as the apostle of the "new science," the man who had tamed lightning (regarded before Franklin's famous kite experiment as a mysterious weapon of divine punishment). Everywhere in Paris there were medallions and banners proclaiming *Eripuit Celeo Fulmen Septrumquie Tyrannis* (He Snatched Lightning from the Sky and the Scepter from Tyrants).

King Louis XVI, however, did not share the popular enthusiasm for the American superstar. To Louis, Franklin represented a dangerous type: a scientist independent of state and king. All well and good to invent lightning rods, but how did that increase the power of the state's cannons or make its fighting ships more powerful? Louis resolved that no French scientist would ever have such independence, frittering away his time in pointless researches that did not enhance the power and the might of France.

Meanwhile, to demonstrate his dislike of Franklin, he gave his favorite mistress a chamber pot containing a medallion of the American minister at the bottom of the bowl. Apprised of this insult, Franklin merely smiled tolerantly, as though he was able to foresee the moment, just a few years distant, when Louis would be taken in a cart before jeering crowds to the guillotine.

A scientific deduction, possibly.

Sources

Introduction: The Ghost in the Machine

Masada: Shaye J. D. Cohen, "Masada," *Journal of Jewish Studies* (Spring–Autumn 1982). Cohen casts strong doubt on Josephus' account of the Masada defenders committing mass suicide, an account he concludes was invented for certain political reasons, chiefly a Roman attempt to justify its brutal suppression of Judean dissidents by portraying the Zealots as unreasoning fanatics willing to murder their own children. More likely, he concludes, the Romans slaughtered all the Zealots once they conquered Masada, citing the archaeological evidence of piles of skeletons found in caves in the face of the plateau—where the Romans probably threw the bodies of their victims.

Roman military power: Michael Grant, *The Army of the Caesars* (New York, 1984).

Ballista: E. W. Marsden, *Greek and Roman Artillery* (Oxford, 1969).

Early war: Lawrence H. Keeley, *War Before Civilization* (London, 1996).

Beginning of civilization: J. G. Cowlett, *Ascent to Civilization* (London, 1984).

Evolution of war: Jacob Bronowski, *The Ascent of Man* (Boston, 1973).

268 years of peace: Will and Ariel Durant, *The Lessons of History* (New York, 1968).

Hobbes's conclusion: Thomas Hobbes, *The Leviathan* (New York, 1946).

Science and war: A. Rupert Hall, "Science, Technology, and Warfare," paper presented at the U.S. Air Force Third Military History Symposium, Denver, Colorado (September 13, 1969).

Feynman: Richard P. Feynman, "The Uncertainty of Science," John Danz Lecture, University of Washington (April 23, 1963).

Shamans: Harry H. Turney-High, *Primitive War* (Columbia, S.C. 1949).

Babylonian, Chinese shamans: Howard Bloom, *The Lucifer Principle* (New York, 1995).

1: "The Valor of Men Is Ended!"

Egyptian infantryman: Miriam Lichtheim, ed., *Ancient Egyptian Literature* (Berkeley, Calif., 1973).

Chariots: Yigal Yadin, *Warfare in Biblical Lands in the Light of Archaeological Study* (New York, 1963).

Composite bow: Jac Weller, *Weapons and Tactics* (New York, 1973).

Ancient Near East: V. Gordon Childe, *The Most Ancient Near East* (London, 1928).

Bronze: J. F. C. Fuller, *Armament and History* (New York, 1945).

Chariot invasion: George Raudzens, "War-Winning Weapons: The Measurement of Technological Determinism in Military History," *Journal of Military History* (October 1990).

Chariot aristocracy: Keith C. Steele and George Steindorf, *When Egypt Ruled the East* (Chicago, 1957).

Kadesh: Gene H. Dickens, "The Armies of Kadesh," *Command* (November–December 1990).

Egyptian complacency: G. Elliott Smith, *The Ancient Egyptians and the Origin of Civilization* (London, 1923).

Iron: Trevor N. Dupuy, *The Evolution of Weapons and Warfare* (New York, 1980).

Triumph of iron armies: Robert Drews, *The End of the Bronze Age* (Princeton, N.J., 1993).

New warship: R. J. Forbes, *Studies in Ancient Technology* (Leiden, 1958).

Assyrian military machine: J. K. Anderson, *Military Theory in the Age of Xenophon* (Berkeley, Calif., 1970).

Helepolis: L. Sprague de Camp, *The Ancient Engineers* (New York, 1960).

Assyrian battering ram machine: Erich F. Schmidt, *Persepolis* (Chicago, 1953).

"I cut off their heads": James Pritchard, *Ancient Near Eastern Texts* (Princeton, N.J., 1955).

Decline of Assyrian military power: Richard Humble, *Warfare in the Ancient World* (London, 1980).

Pre-Socratic philosophers: David D. Lindley, *The Beginning of Western Science* (Chicago, 1992).

Plato, Aristotle, Thucydides: Donald Kagan, *On the Origins of War and the Preservation of Peace* (New York, 1995).

Thales of Melitus: G. E. R. Lloyd, *Early Greek Science: Thales to Aristotle* (London, 1970).

Marathon: F. E. Adcock, *The Greek and Macedonian Art of War* (Berkeley, Calif., 1957). According to Greek accounts of the battle, a messenger was sent back to Athens with news of the great victory. He ran the twenty-six miles nonstop, gasped out the news, then dropped dead of exhaustion—an event commemorated ever since in the Olympics by the marathon race.

Greek approach to science: W. P. D. Wightman, *The Growth of Scientific Ideas* (New Haven, Conn., 1953).

Troy: J. F. C. Fuller, *A Military History of the Western World*, vol. 1 (New York, 1954).

Torsion artillery: Ralph Payne-Gallwey, *A Summary of the History, Construction, and Effects in Warfare of the Projectile-Throwing Engines of the Ancients with a Treatise on the Structure, Power, and Management of Turkish and Other Oriental Bows of Medieval and Later Times* (London, 1907).

Science of torsion artillery: A. M. Snodgrass, *Arms and Armor of the Greeks* (New York, 1967).

Machines' influence on Greek science: Werner Soedel and Vernard Foley, "Ancient Catapults," *Scientific American* (March 1979).

Siege machine: John Warry, *Warfare in the Classical World* (London, 1980).

"Oh, Heracles . . .": Quoted in Sidney Toy, *A History of Fortification from 3,000 B.C. to A.D. 1700* (London, 1955).

Siege of Rhodes: A. G. Drachmann, *The Mechanical Technology of Greek and Roman Antiquity* (Copenhagen, 1963).

Philip of Macedon, Alexander the Great: W. W. Tarn, *Alexander the Great* (London, 1948).

Greek weapons: George Sarton, *Ancient Science Through the Golden Age of Greece* (New York, 1993).

Alexandria Museion: E. A. Parsons, *The Alexandrian Library* (London, 1952).

Science breakthroughs at Alexandria: Marshall Clagett, *Greek Science in Antiquity* (New York, 1955).

Pure vs. applied science: G. E. R. Lloyd, *Early Greek Science* (New York, 1970).

Archimedes: Paul Hoffman, *Archimedes' Revenge* (New York, 1987).

Archimedes at Syracuse: Polybius, *The Histories of Polybius* (London, 1889).

Roman Empire: Arnold Toynbee, *War and Civilization* (New York, 1950).

Roman attitude toward science: Toby E. Huff, *The Rise of Early Modern Science* (Cambridge, Eng., 1993).

Julius Caesar: Hans Delbruck, *History of the Art of War* (New York, 1981).

Roman disdain for science: Charles Singer, *A History of Technology* (Oxford, 1957).

"No Roman . . .": Quoted in V. Gordon Childe, *Man Makes Himself* (New York, 1951).

Carthaginian warships: Honor Frost, "How Carthage Lost the Sea," *Natural History* (December 1987).

Barbarians: Theodor Mommsen, *The History of Rome* (London, 1911).

Scythians: David Quammen, "The Ineffable Union of Horse and Man," in Robert Cowley, ed., *Experience of War* (New York, 1992).

Barbarian invasion: Arnold Pacey, *Technology in World Civilization* (Cambridge, Eng., 1991).

Battle of Adrianople: Wlodzimierz Onacewicz, "Armored Cavalry Systems, 378–1346 A.D.," *History, Numbers and War* (June 1980).

Fall of Rome: Ferdinand Lott, *The End of the Ancient World and the Beginnings of the Middle Ages* (New York, 1931).

2: Bride of Faith

Agincourt: Albert Nofi, "Agincourt: The Triumph of Archery over Armor, 25 October 1415," *Strategy and Tactics* (May–June 1981).

Origin of longbow: Ewart Oakeshott, *The Archaeology of Weapons* (New York, 1994).

Development of longbow: Robert Hardy, *Longbow* (Cambridge, Eng., 1976).

Edward I, Edward II: Vic Hurley, *Arrows Against Steel* (New York, 1975).

Dark Ages: Hermann Kesten, *Copernicus and His World* (New York, 1945).

"It is the last hour": Quoted in J. F. C. Fuller, *Armament and History* (New York, 1945).

Armored cavalry: Archer Jones, *The Art of War in the Western World* (Chicago, 1987).

Martel, Charlemagne: Lynn White, *Medieval Technology and Social Change* (Oxford, 1962).

Knights: J. F. Verbruggen, *The Art of Warfare in Western Europe During the Middle Ages* (New York, 1977)

Hastings: Richard Barber, *The Knight and Chivalry* (London, 1970).

Augsburg armorers: James G. Mann, "Notes on the Evolution of Plate Armor in Germany in the Fourteenth and Fifteenth Centuries," *Archaeologia* LXXIV (1935).

Crossbow: Robert L. O'Connell, "The Life and Hard Times of the Crossbow," in Robert Cowley, ed., *Experience of War* (New York, 1992).

Castles: Armin Tuulse, *Castles of the Western World* (Vienna, 1958).

Byzantium: George Ostrogorsky, *History of the Byzantine State* (New Brunswick, N.J., 1969).

Greek fire: Cecil Torr, *Ancient Ships* (Cambridge, 1895).

Resurgence of Islam: Stanford J. Shaw, *Empire of the Gazis: The Rise and Decline of the Ottoman Empire, 1280–1808* (Cambridge, Eng., 1976).

King Alfonso: Marshall Claggett et al., eds., *Twelfth-Century Europe and the Foundations of Modern Society* (Madison, Wis., 1961).

Toledo: Thomas Goldstein, *Dawn of Modern Science* (Boston, 1988).

Treasures of Toledo: D. M. Dunlop, *Arabic Science in the West* (Karachi, 1958).

Greek texts: Thomas S. Kuhn, *The Structure of Scientific Revolutions* (Chicago, 1970).

Thomas Aquinas: David D. Lindberg, *The Beginnings of Western Science* (Chicago, 1992).

"God cannot . . .": Quoted in J. W. Baldwin, *The Scholastic Culture of the Middle Ages* (Lexington, Ky., 1971).

Roger Bacon: J. H. Bridges, *The Life and Work of Roger Bacon* (London, 1914).

Crusaders in Constantinople: Charles Oman, *A History of the Art of War in the Middle Ages* (London, 1924).

Gutenberg: Elizabeth L. Eisenstein, *The Printing Press as an Agent of Change* (Cambridge, Eng., 1979).

Flavius Vegetius Renatus: Charles R. Shrader, "The Influence of Vegetius' *De Rei Militari*," *Military Affairs* (December 1981); Flavius Vegetius Renatus, *Military Institutions of the Romans* (Harrisburg, Pa., 1941).

Aldus Manutius: John Herman Randall, *The Making of the Modern Mind* (New York, 1926).

Military best-sellers: Lucien Febvre and Henri-Jean Martin, *The Coming of the Book* (London, 1976).

Beginnings of the military revolution: Frank Tallet, *War and Society in Early-Modern Europe, 1495–1715* (London, 1952).

Ghazzali: T. W. Arnold, *The Legacy of Islam* (Oxford, 1931).

Muslim science: Seyged Hossein, *Science and Civilization in Islam* (New York, 1992).

Bacon and gunpowder: W. L. Hine, *Gunpowder and Ammunition: Their Origin and Progress* (London, 1904).

Montigny: Thomas Esper, "The Replacement of the Longbow by Firearms in the English Army," *Technology and Culture* 6 (1965).

3: The Dragon's Teeth

Fall of Monte San Giovanni: F. L. Taylor, *The Art of War in Italy, 1494–1529* (London, 1921).

Panic in Italy: Robert L. O'Connell, *Of Arms and Men* (Oxford, 1989).

Dürer, Michelangelo: Lewis Mumford, *Technics and Civilization* (New York, 1962).

Leonardo da Vinci: Antonia Vallentin, *Leonardo da Vinci* (New York, 1938).

"Awoke too early . . ." Quoted in Sherwin Nuland, *Leonardo da Vinci* (New York, 2000).

da Vinci's notebooks: Edward McCurdy, ed., *The Notebooks of Leonardo da Vinci* (New York, 1939).

Charles VII, king of France: Geoffrey Parker, *The Military Revolution* (Cambridge, Eng., 1988).

Chinese "weapons of fire": Carlo M. Cipolla, *Guns, Sails, and Empires* (London, 1965).

Arab *madfaa:* Henry W. L. Hine, *The Origin of Artillery* (New York, 1915).

Early culverins, bombards: J. R. Parrington, *A History of Greek Fire and Gunpowder* (New York, 1960).

French development of cannons: Friedrich Klemm, *A History of Western Technology* (New York, 1956).

Trace Italienne: William H. McNeil, *The Pursuit of Power* (Chicago, 1982); J. R. Hale, *Europe in the Late Middle Ages* (Evanston, Ill., 1965).

Trigonometry and the *trace Italienne:* Ian Stewart, *The Magical Maze* (London, 1998).

Revolutionary effect of new weapons: Felix Gilbert, "Machiavelli: The Renaissance of the Art of War," in Edward Mead Earle, ed., *Makers of Modern Strategy* (Princeton, N.J., 1943).

Renaissance of science: W. E. Lecky, *History of the Rise and Influence of the Spirit of Rationalism in Europe* (London, 1910).

Disinterest in theoretical science: Jacob Burckhardt, *The Civilization of the Renaissance in Italy* (London, 1974).

Obsession with astrology: Ingrid D. Rowland, "Star Trek," *New York Review of Books* (February 22, 2001).

Alchemy becomes chemistry: F. J. Moore, *A History of Chemistry* (New York, 1918).

Niccolò Tartaglia: Marshall Clagett, *The Science of Mathematics in the Middle Ages* (London, 1959).

Problem of ballistics: A. Rupert Hall, *Ballistics in the Seventeenth Century* (Cambridge, Eng., 1952).

"It was a thing . . .": Quoted in S. R. Bordo, *The Flight to Objectivity* (Albany, N.Y., 1989).

"The wolf . . .": Quoted in W. W. Rouse Ball, *A Short Account of the History of Mathematics* (London, 1908).

Galileo: Stillman Drake, *Galileo* (Oxford, 1980).

Galileo's "geometric and military compass": Hugh Kearney, *Science and Change, 1500–1700* (New York, 1971).

Galileo and the Medicis: G. F. Young, *The Medicis* (New York, 1909).

Venice arsenal: Richard W. Unger, *The Ship in the Medieval Economy* (Montréal, 1980).

Galileo's telescope: James Burke, "A Few Notes," *Scientific American* (July 1999).

Satellites of Jupiter: Dava Sobel, *Galileo's Daughter* (New York, 1999).

Spanish *arquebus:* J. R. Hale, "Gunpowder and the Renaissance," in C. H. Carter, ed., *From Renaissance to Counterreformation* (New York, 1966).

Spanish musket: Christopher J. Duffy, *The Military Experience in the Age of Reason* (New York, 1988).

Battle of Pavia: B. P. Hughes, *Firepower: Weapons Effectiveness on the Battlefield* (New York, 1975); R. J. Knecht, *Francis I* (London, 1982).

Islamic science: V. J. Parry and M. E. Yapp, eds., *War, Technology, and Society in the Middle East* (London, 1975).

Empress Tsin Jeng: Joseph Needham, *Science and Civilization in China* (Cambridge, Eng., 1954).

"No more writing . . .": Quoted in Robert Temple, *The Genius of China: 5,000 Years of Science, Discovery, and Invention* (New York, 1987). When Marco Polo visited the mighty Mongol Empire in the early thirteenth century, he noted that Genghis Khan had five thousand astrologers and soothsayers on hand to advise him on the most propitious moments to take actions large and small. However, the Mongol leader had no scientific advisers, whose time would have been wasted in any event: Their ruler was totally illiterate and had no interest in even the basic natural phenomena—save a large number of concubines. The Mongols would pay dearly for this scientific disinterest: Their vaunted light cavalry, which terrified much of the civilized world for several centuries, at one point threatening to overcome Europe, was never improved, and finally was defeated by the power of superior European military technology. See Bradley E. Schaefer, "Conjunctions That Changed History," *Sky and Telescope* (May 2000).

Constriction of Chinese science: A. P. Usher, *History of Mechanical Invention* (New York, 1929).

Matteo Ricci: Jonathan D. Spence, *The Memory Palace of Matteo Ricci* (New York, 1999).

1629 solar eclipse: William Boulting, *Four Pilgrims* (New York, 1920).

4: Outward Bound

Battle of Cajamarca: John Hemmings, *The Conquest of the Incas* (San Diego, 1970).

"We Europeans . . .": Quoted in George Bailey, *Galileo's Children* (New York, 1988).

Christopher Columbus: Samuel Eliot Morison, *Admiral of the Ocean Sea* (Boston, 1942).

Fall of Ceuta: Alfred W. Crosby, *Biological Imperialism: The Biological Expansion of Europe, 900–1900* (Cambridge, Eng., 1986).

Prince Henry: Peter Russell, *Prince Henry the Navigator* (New Haven, Conn., 2000).

European obsession with spices: Carlo M. Cipolla, *European Culture and Overseas Expansion* (London, 1970).

Navigation problems: Thomas Goldstein, "Geography in Fifteenth-Century Florence," in John Parker, ed., *Merchants and Scholars* (Minneapolis, 1965).

Sagres Harbor: C. R. Beazley, *Prince Henry the Navigator* (London, 1901).

Caravelle **design:** J. F. Guilmartin Jr., *Gunpowder and Galleys* (Cambridge, Eng., 1974).

Beginning of the slave trade: Armando da Silva Saturnino Monteiro, "The Decline and Fall of Portuguese Sea Power, 1583–1663," *The Journal of Military History* (January 2001).

Columbus and imperialism: Joseph Judge et al., "Columbus and the New World," *National Geographic* (November 1986).

Compass, navigation difficulties: C. R. Boxer, *The Portuguese Seaborne Empire* (London, 1969).

Paolo Toscanelli: Dana B. Durand, "Tradition and Innovation in Fifteenth Century Italy," *Journal of the History of Ideas* (January 1943).

Henry VIII and naval development: Frederick Leslie Robertson, *The Evolution of Naval Armament* (London, 1921).

Vasco da Gama's voyage: Luc Cuyvers, *Into the Rising Sun: Vasco da Gama and the Search for a Sea Route to the East* (London, 1999).

Great Harry: Peter Padfield, *Guns at Sea* (New York, 1973).

William Gilbert: Duane H. Roller, *The de Magnete of William Gilbert* (Amsterdam, 1959).

Drake's voyage: Kenneth R. Andrews, *Elizabeth Privateering, 1585–1603* (Cambridge, 1964).

English privateers, Spanish Armada: Garrett Mattingly, *The Armada* (New York, 1980).

New English ships: Boies Penrose, *Travel and Discovery in the Renaissance* (Cambridge, 1952).

Ne plus ultra: J. D. Bernal, *Science in History* (London, 1957).

Bull of Demarcation: Rupert T. Gold, *The Marine Chronometer* (London, 1923).

Tycho Brahe: J. A. Goode, *Life and Times of Tycho Brahe* (Princeton, 1947).

Johannes Kepler: Arthur Koestler, *The Watershed* (Lanham, Md., 1960).

Latitude, longitude: Lloyd A. Brown, *The Story of Maps* (Boston, 1949).

Royal Observatory: J. G. Burke, ed., *The Uses of Science in the Age of Newton* (Berkeley, Calif., 1983).

Paris Observatory: Derek House, *Greenwich Time and the Discovery of the Longitude* (New York, 1980).

Competing zero meridians: Lisa Jardine, *Ingenious Pursuits* (New York, 1999).

Gian Cassini: Harcourt Brown, *Scientific Organizations in Seventeenth-Century France* (Baltimore, 1934).

John Flamsteed: Charles Singer et al., *A History of Technology* (Oxford, 1958).

Harrison clock: Dava Sobel, *Longitude* (New York, 1998).

5: The Final Argument of Kings

Incident at Odenaarde: J. L. Motley, *The Rise of the Dutch Republic,* vol. 3 (New York, 1856).

Sixteenth-century warfare: André Corvisier, *Armies and Societies in Europe, 1494–1789* (Bloomington, Ind., 1979).

Ultimo ratio regis: Martin van Creveld, *Technology and War* (New York, 1972).

Change in the conduct of war: Geoffrey Parker, *The Military Revolution* (Cambridge, Eng., 1988).

"This is the century . . .": Quoted in J. J. Brown and J. H. Elliott, *A Palace for a King* (New Haven, Conn., 1980).

Increasing scope and violence of warfare: J. S. Levy, *War in the Great Power System* (Lexington, Ky., 1983).

Golden age of science and war: George Clark, *War and Society in the Seventeenth Century* (London, 1958).

Components of the scientific revolution: Michael Roberts, *The Military Revolution, 1560–1660* (Belfast, N. Ire., 1956).

Prince Maurice of Nassau: P. J. Davis and Reuben Hersh, *Descartes' Dream* (Boston, 1986).

Simon Stevin: Edward Jan Dijsterhuis, *Simon Stevin* (The Hague, 1970).

Stevin's superbomb: Edward Grierson, *The Fatal Inheritance* (Garden City, N.Y., 1969).

Innovations of Maurice and Stevin: Library of Congress, *The Evolution of International Technology* (Washington, D.C., 1970).

Louis XIV: John Dalberg-Acton (Lord Acton), *Lectures on Modern History* (London, 1950).

Louis XIV and science: Penfield Roberts, *The Quest for Security* (New York, 1947).

Jean-Baptiste Colbert: Charles Cole, *Colbert and a Century of French Mercantilism* (Hamden, Conn., 1964).

Advances in French cartography: Arthur Berry, *A Short History of Astronomy* (New York, 1909).

"Excellent, gentlemen . . .": Quoted in Voltaire, *Age of Louis XIV* (London, 1975).

Scientific advances: John U. Nef, "War and Economic Progress, 1540–1740," *Economic History Review* 12 (1942).

Gunpowder: William T. de Bary, *Message of the Mind* (New York, 1989).

Beginning of industrial revolution: Lewis Mumford, *Technics and Civilization* (New York, 1961).

Pierre Beauchamp: James Burke, "Turkish Delight," *Scientific American* (March 1998).

Picard at Versailles: Henri Markton, *The Age of Louis XIV* (Boston, 1865).

Royal Society for Improving Natural Knowledge: Alfred North Whitehead, *Science and the Modern World* (New York, 1926).

French, Prussian militaries: Michael Howard, *Studies in War and Peace* (London, 1970).

John Napier: Bernard Brodie and Fawn M. Brodie, *From Crossbow to H-Bomb* (New York, 1960).

"Too many devices . . .": Quoted in John U. Nef, *War and Human Progress* (New York, 1968).

Benjamin Robins: Breit David Steele, "The Ballistics Revolution: Military and Scientific Change from Robins to Napoleon," (Ph.D. diss., University of Minnesota, 1994).

Carronade: Ian Hogg, *Artillery* (New York, 1973).

Rifled barrels, breechloaders: W. Y. Carman, *A History of Firearms* (London, 1955).

Gunpowder factories: Peter N. Stearns, *The Industrial Revolution in World History* (Boulder, Col., 1993).

Honoré de Blanc: L. T. C. Rolt, *Tools for the Job* (London, 1965). The ever-alert Thomas Jefferson, who had seized upon the decimalization idea for coinage, heard about de Blanc's contretemps and instantly realized the Frenchman's idea was just the ticket for the American revolutionaries, who would need a lot of cannons and muskets. He informed George Washington, who in 1798 gave a contract to inventor Eli Whitney to create a centralized manufacturing process for cannons and muskets. Whitney developed the first industrialized production process in America, ultimately known as the "American system." Featuring the centralization of all aspects of a manufactured item under one roof and production on a systematic, assembly-line basis, it was later extended to the manufacture of locomotives, bicycles, sewing machines, and automobiles, making the United States an industrial giant.

French cannon crisis: D. S. L. Cardwell, *Steam Power in the Eighteenth Century* (London, 1963).

John Wilkinson: Ron Davies, *John Wilkinson* (London, 1987).

Wilkinson and Watt: Carlo M. Cipolla, ed., *The Industrial Revolution, 1700–1914* (Brighton, Eng., 1976).

British industrial revolution: Clive Trebilcock, "Spin-off in British Economic History: Armament and Industry, 1760–1914," *Economic History Review* 22 (1969).

Chappe, British communications systems: James Burke, "Cheers," *Scientific American* (May 1998).

6: Prometheus Unchained

Society for the Encouragement of French Industry: "Canned History," *New Scientist* (November 28, 1998).

Nicholas Appert's invention: Michael Howard, *War in European History* (Oxford, 1976).

Peter Durand: Harold R. L. Williamson, "The Secret of Napoleon's Mobility," *Military Science and Technology* (August 1979). Appert and Durand unwittingly would make a tremendous contribution to medical science. They knew that boiling food and then sealing it in an airtight container prevented putrefaction, but didn't realize that the process killed germs, the cause of disease. Many years later, a French scientist named Louis Pasteur would take another look at Appert's sealed bottle and realize he was on to something: boiling killed germs.

Antoine Lavoisier: Arthur Donovan, *Antoine Lavoisier* (Oxford, 1993).

Lavoisier's contributions to French military power: Sidney French, *Torch and Crucible: The Life and Death of Antoine Lavoisier* (Princeton, N.J., 1941). Lavoisier also played a critical but unacknowledged role in the American Revolution. When France decided to support the American rebels, Lavoisier was assigned a top-secret task: dramatically increase production of French gunpowder so that large supplies could be covertly shipped to America, where the colonists had no access to large-scale production of gunpowder. Lavoisier succeeded, and in 1789 joked to friends that the Americans could attribute their defeat of the British Empire to three factors: George Washington, the Continental soldier, and French gunpowder.

"The revolution has no need...": Quoted in David Mckie, *Antoine Lavoisier* (New York, 1952).

Gaspard Monge: Hans Speier, "Militarism in the Eighteenth Century," *Social Research* III (1936).

New French cannons: B. P. Hughes, *Firepower: Weapons' Effectiveness on the Battlefield, 1630–1850* (London, 1974).

Napoleon's destruction of Egyptian, Prussian armies: David Chandler, *The Campaigns of Napoleon* (New York, 1971).

Napoleon and science: Ken Adler, *Engineering the Revolution* (Princeton, N.J., 1997).

Oxidation of ammonia: Glenys Crocker, *The Gunpowder Industry* (Haverfordwest, Eng., 1986).

Jacques Conte and the Montgolfier brothers: L. T. C. Rolt, *The Aeronauts* (Gloucestshire, Eng., 1985).

Fulton and Napoleon: Bernard Brodie, *Sea Power in the Machine Age* (Princeton, N.J., 1942).

Prussia reorganizes: Gordon A. Craig, *The Politics of the Prussian Army, 1640–1945* (Oxford, 1955).

Prussian Kriegsakademie: Peter F. Hauser, "Lessons from the Kriegsakademie," *Airpower Journal* (Autumn 1997).

University of Berlin: Karl R. Popper, *The Open Society and Its Enemies* (Princeton, N.J., 1962).

German science education system: I. Bernard Cohen, *Revolution in Science* (Cambridge, Eng., 1985).

"It brings army...": Quoted in Paul M. Kennedy, *The Rise and Fall of Great Powers* (New Haven, Conn., 1982).

Kaiser Wilhelm Institute: Lothar Burchardt, *The Politics of Science in Wilhelmian Germany* (Göttingen, 1975).

British loss of lead in science and technology: Harry Edward Meal, *From Spinning Wheel to Spacecraft* (New York, 1965).

Rise of Japan: Jean-Marie Bonthous, "The Japanese Approach to Intelligence," *The Journal of Intelligence and Counterintelligence* 7, no. 5 (1995).

War against China, Battle of Tsushima: Joseph Barnes, *Empire in the East* (New York, 1934).

Peter the Great and the rise of Russian military power: Gordon Turner, *A History of Military Affairs in Western Society since the Eighteenth Century* (New York, 1953).

Gun cotton: James Burke, *The Pinball Effect* (New York, 1996).

Alfred Nobel and dynamite: Erik Bergengren, *Alfred Nobel* (London, 1962).

Napoleon III's prize: A. W. Wilson, *The Story of the Gun* (London, 1944).

Bessemer process: Jean Colin, *The Transformation of War* (London, 1912).

Alfred Krupp: Peter Batty, *The House of Krupp* (London, 1966).

British shell design: Michael Howard, ed., *The Lessons of History* (New Haven, Conn., 1991).

Monitor, Merrimac: J. P. Baxter, *The Introduction of the Ironclad Warship* (Cambridge, Eng., 1933).

John Ericsson: Robert V. Bruce, *The Launching of Modern American Science, 1846–1876* (New York, 1987).

HMS *Dreadnought:* John Tetsuro Sumida, "Sir John Fisher and the *Dreadnought*," *Journal of Military History* (October 1995).

John P. Holland and his submarine: Richard Knowles Morris, *John P. Holland* (Annapolis, Md., 1966).

Advances in artillery: Trevor N. Dupuy, *Weapons and Warfare* (New York, 1968).

Machine guns, Hiram Maxim, Omdurman: John Ellis, *The Social History of the Machine Gun* (New York, 1975).

Militarization of Europe pre–World War I: Dennis Showalter, *Railroads and Rifles* (New York, 1977).

Discoveries of Fritz Haber: Andrew Clow, *The Chemical Revolution* (London, 1952).

Fritz Haber: Fritz Stern, "Fritz Haber: The Scientist in Power and in Exile," lecture at the seventy-fifth anniversary celebration of the Fritz Haber Institute for Physical Chemistry and Electrochemistry, Berlin (October 11, 1986). After World War I, Haber helped found the Emergency Association of German Science, which sought to raise funds to continue science research in a war-devastated Germany. As an indication of how little he had learned from the disaster of World War I, Haber justified the work of the association this way: "Our existence as a people depends on the maintenance of our intellectual great power position, which is inseparable from our scientific enterprise." Haber went on to continue work at his institute until 1933, when the Nazis began to exert strong pressure on him to fire all Jewish scientists. Haber, who was Jewish himself, refused, and bravely told his tormentors that he hired scientists strictly on the basis of their scientific capabilities, "not who their grandmothers happened to be." But a year later, he was driven out of Germany after the

Nazis called this man, one of Germany's greatest scientists, "a Jew swine." Haber decided to immigrate to Palestine but died en route, puzzled by yet another course of events his brilliant scientific mind was incapable of understanding.

Poison gas: Allan D. Beyerchen, *Scientists Under Hitler* (New Haven, Conn., 1977).

Haldane's defense of Haber: J. B. S. Haldane, *Callincus: A Defense of Chemical Warfare* (London, 1924).

7: The Sorcerer's Apprentices

British secret mission: David Zimmermann, *Top Secret Exchange* (Montréal, 1996).

Vannevar Bush's proposal: Philip H. Adebon, "The President's Science Advisers," *Minerva* (Winter 1965).

Office of Scientific Research and Development recruitment: Vannevar Bush, *Modern Arms and Free Men* (New York, 1949).

National Research Council, World War I: Hunter Dupree, *Science in the Federal Government* (Cambridge, Mass., 1957).

British Admiralty and German U-boat menace: E. J. Marder, *From Dreadnought to Scapa Flow*, vol. 2 (London, 1962).

Tanks: B. H. Liddell Hart, *The Tanks* (London, 1949). The term "tank" came about because British scientists and technicians working on the project, in an attempt to keep it secret, spread the story that they were working on developing cisterns for the Russian military. The basic idea for the tank came not from science but from a California farmer. Facing a shortage of horses (draft animals were taken by the government for war) and working on soggy ground, he came up with the ingenious idea of wide, continuous segmented tracks that spread the weight of the vehicle and provided tremendous traction. They proved ideal for transportation across the muddy fields of the Western Front, and it was but a short step to realize that the system would be perfect for an armored vehicle equipped with guns.

Lindermann: J. F. C. Fuller, *Machine Warfare* (Washington, D.C., 1943).

Ernest Rutherford's complaint: James B. Conant, *Harvard to Hiroshima* (New York, 1993).

Tots and Quots, Zuckerman, Snow: Andrew Sinclair, *The Red and the Blue* (London, 1983).

"The war is compressing . . .": "Rapid Development of Science," *London Times* (November 9, 1942).

Development of airships: Basil Collier, *A History of Air Power* (New York, 1974).

Wright brothers and science: Michael Guillen, *Five Equations That Changed the World* (New York, 1995).

Military implications of Wright brothers' plane: Marvin W. McFarland, "When the Airplane Was a Military Secret," *Air Power Historian* II (1955).

Impact of airplane in Europe: Phil Scott, "What Were They Thinking?" *Air and Space* (March 2001).

Libya 1911: Robin Higham, *Airpower* (New York, 1972).

German Gotha bomber: Robert L. O'Connell, "The Gotha and the Origins of Strategic Bombing," *Military History Quarterly* (Autumn 1998).

Role of science and technology in the development of air weapons: Edward W. Constant, *The Origins of the Turbojet Revolution* (Baltimore, 1980).

Solid-state transistors, ballpoint pen: S. A. Leinwoll, *From Spark to Satellite* (New York, 1979).

G. H. Hardy: Solomon W. Golomb, "Mathematics Forty Years after Sputnik," *American Scholar* (Spring 1998).

Development of cipher machines: David Kahn, *The Codebreakers* (New York, 1967).

Alan Turing: Andrew Hodges, *Alan Turing: the Enigma* (New York, 1983). Despite his incalculable contribution to the war effort, Turing's life ended tragically and shabbily at the hands of the British government. Turing was a flamboyant homosexual at a time when homosexuality was a serious crime in Britain. In 1952 he was arrested on sodomy charges and was offered the choice of prison or chemical treatments that supposedly would end his homosexuality. Turing opted for the treatments, which among other horrible side effects caused his breasts to grow to large size. In 1953 he committed suicide with cyanide he had prepared himself and injected into an apple. He left an estate of less than $18,000, and his funeral was attended by only three people, including his mother and brother. The British government never conferred any honors on him, nor does there exist any monument to his memory. Because of stringent secrecy regulations, the ULTRA operation was not revealed until 1974, when for the first time the world of science learned that the oddball Cambridge mathematician was the father of the remarkable machine that is still revolutionizing human existence.

Turing's *bombe:* Joseph Garlinski, *The Enigma War* (London, 1978).

ENIAC: Martin H. Weik, "The ENIAC Story," *Ordnance* (January–February 1961).

Mark I: Kathleen B. Williams, "Scientists in Uniform," *Naval War College Review* (Summer 1999).

Post–World War II transistor, integrated circuit: Keith Ferrell, "Electronic Worlds Without End," *Omni* (October 1993).

Konrad Zuse computer: James W. Cortada, *Before the Computer* (Princeton, N.J., 1993). Zuse was unaware that the SS at the same time was using an IBM card-punch system to collate data on Jews, a preparatory effort for the Final Solution. See Edwin Black, *IBM and the Holocaust* (New York, 2001).

Fragmented German research and development effort: Richard P. Hallion, "Doctrine, Technology, and Air Warfare," *Airpower Journal* (Fall 1987). German science also was severely handicapped by its Nazification, which drove out many of its greatest scientists because they happened to be Jews. Between 1933 and 1938 a total of 1,880 first-class scientists were forced out of their university posts, including 25 percent of Germany's Nobel Prize winners. Joseph Needham, *The Nazi Attack on International Science* (London, 1941).

German long-range bomber: U.S. Army Ordnance Department Ballistic Research Laboratory, *Report on German Scientific Establishments* (Aberdeen, Md., 1947).

Problems of early rockets: Kenneth H. Roper, "Rocket Propulsion for Space," *Air University Review* (Winter 1966).

William Congreve's rockets: Alfred Price, *The Guided Weapons* (London, 1971).

Robert Goddard: Carl Sagan, commencement address, Clark University, Worcester, Mass. (May 18, 1978).

German rocket enthusiasts, Wernher von Braun: Michael J. Neufeld, *The Rocket and the Reich* (New York, 1995). The German army's specific interest in von Braun's rockets stemmed from an oversight in the post–World War I Versailles Treaty, which strictly limited the military technologies Germany was permitted to research or develop. The treaty made no mention of rockets, since military experts at the time could not foresee any significant military use for them.

V-2 rocket: Brian J. Ford, *The Rocket Race* (London, 1971).

Fritz-X guided missile: Charles H. Bogart, "German Remotely Piloted Bombs," *U.S. Naval Institute Proceedings* (November 1976). The Fritz-X's radio guidance signal was controlled by a lever that coordinated the signal and the flight path of the missile. German pilots nicknamed the lever "joystick," a name that stuck when the system was used many years later for control of video and computer games.

Chaff: D. E. Gordon, *Electronic Warfare* (London, 1981).

Advanced depth charge: Kenneth Poolman, *The Winning Edge: Naval Technology in Action, 1939–1945* (Annapolis, Md., 1997).

Bazooka: Elting E. Morison, *Men, Machines, and Modern Times* (Cambridge, Mass., 1966).

Antimine technology: Dan van der Vat, *Stealth at Sea* (London, 1994).

P-51 Mustang: Ray Wagner, *Mustang Designer: Edgar Schmuel and the P-51* (Washington, D.C., 2000). The P-51 represented a triumph of British and American science and technology, a partnership that produced an airframe incorporating advanced American metallurgy; a "bubble" canopy of a new, bulletproof plastic called Plexiglas; an advanced British Rolls-Royce engine known as Merlin; and innovative "drop tanks" of extra fuel that allowed the Mustangs to escort bombers all the way into Germany, tanks that could be released when the P-51 pilots wanted lighter weight to take on German fighters. Developed specifically to protect unescorted bombers that were suffering heavy losses to Luftwaffe fighters, the P-51 was far advanced technologically over any German fighter and soon swept the Luftwaffe from the skies.

Graf Zeppelin II **espionage mission:** William Scheck, "Luftschiff Zeppelin 130," *Command* (July–August 1993). Radar, actually an acronym for "radio detection and ranging," illustrates how even the most non-military of scientific developments can have significant military applications. Radar, which originally grew out of research into radio atmospherics, was developed to locate storms by means of using radio signals to detect electrically charged clouds, the certain prelude to thunderstorms. But British scientists discovered, to their surprise, that the signals bounced back from the ionized layer of the atmosphere and that the time interval between emission and return of the signal could be accurately measured, providing the precise distance between the antenna and whatever it had detected. It required only a short step to realize that the same system could detect solid objects in the air (such as airplanes), and that measuring the time between signal and detection would provide direction and range of the detected object. See W. R. Machlaurin, *Invention and Innovation in Radio* (New York, 1949).

British scientific intelligence, R. V. Jones: R. V. Jones, *The Wizard War* (New York, 1976).

OSS scientific intelligence: R. Bradley Smith, *OSS* (Berkeley, Calif., 1969).

Scope of wartime science: Philip M. Morse and George E. Kimball, *Methods of Operations Research* (New York, 1951).

Modern effects of wartime scientific research: H. F. Mark, *Giant Molecules* (New York, 1966).

Nylon, synthetic rubber: C. S. Marvel, "The Development of Polymer Chemistry in America—the Early Days," *Journal of Chemical Education* (July 1981).

Bomb tests on monkeys: C. P. Snow, *Science and Government* (London, 1964).

Napalm: Malvern Lumsden, *Incendiary Weapons* (Cambridge, Eng., 1975).

Hiroshima: Samuel Walker, *Prompt and Utter Destruction* (Chapel Hill, N.C., 1997).

L'Osservatore Romano: John U. Nef, *War and Human Progress* (New York, 1968).

8: A Thousand Suns

Otto Frisch, Lise Meitner: Alan P. Lightman, "To Cleave an Atom," *Science* 84, no. 103 (1984).

Hahn-Strassman experiment: Andrew Brown, *The Neutron and the Bomb* (Oxford, 1997).

Splitting of the atom: Grace Marmor Spruch, "Nobel Tics," *American Scholar* (January 2000).

Niels Bohr at theoretical physics conference: Victor Guillemin, *The Story of Quantum Mechanics* (New York, 1968).

Impact of Hahn-Strassman article: Richard Rhodes, *The Making of the Atomic Bomb* (New York, 1986).

British intelligence source in Germany: Arnold Kramish, *The Griffin* (New York, 1987). The German scientific publisher was Paul Rosbaud, a secret anti-Nazi who became deeply concerned when he learned, via his extensive network of contacts within the German scientific community, of a frightening panoply of secret weapons under development. Recruited by British intelligence in the mid-1930s, he began providing details of weapons development programs to M16, the British foreign intelligence agency that ran an extensive scientific intelligence collection effort under R. V. Jones. Code-named Griffin, Rosbaud throughout the war was an invaluable source who provided the British with details on the V-2 rocket program, the German atomic bomb program, and a long list of other technical and scientific developments. Despite German suspicions that their scientific secrets were leaking out, Rosbaud evaded dectection and managed to survive the war.

Frisch flees to Britain: Richard Peierls, *Atomic Histories* (New York, 1997).

Development of nuclear physics: Helge Kragh, *Quantum Generations* (Princeton, N.J., 1999).

"It is a many-mansioned...": Albert Einstein, *Mein Weltbild* (Zurich, 1953).

Einstein and nuclear physics: Helen Dukas and Banesh Hoffmann, *Albert Einstein: The Human Side* (Princeton, N.J., 1979).

Göttingen: Emilio Segre, *The Physicists* (New York, 1965).

Germany and science: Fritz Stern, "Einstein's Germany," in Gerald Holton and Yehuda Elkana, eds., *Albert Einstein* (New York, 1982).

Einstein and German science: Jeremy Bernstein, *Einstein* (New York, 1973).

Max Planck: James Heilbron, *The Dilemmas of an Upright Man* (New York, 1986).

Nazification of German science: Mark Walker, *Nazi Science* (New York, 1995).

Werner Heisenberg: Martin Gardner, "Werner Heisenberg," *Dimensions* 7, 1 (1993). Scientists in Allied nations assumed that Heisenberg, as Germany's most preeminent nuclear physicist, would be put in charge of that country's atomic bomb development program. That led the American OSS to a harebrained idea: cripple the German program by eliminating Heisenberg, the "great brain" behind it. There was one outright assassination attempt in 1943, when an OSS agent armed with a gun attended a scientific lecture Heisenberg gave in Switzerland. His orders were to shoot Heisenberg if he heard him mention anything like "atomic bomb." Heisenberg didn't, so the agent decided there was no point in shooting him. In 1944, the OSS planned an intricate kidnapping operation, which involved snatching Heisenberg from another scientific conference in Switzerland, spiriting him out of the country by plane, transferring him to a ship somewhere in the Mediterranean, then taking him to the United States, where he would tell what he knew. Cooler heads finally prevailed to cancel the mad scheme; apparently, those planning the operation didn't bother to consider such things as how the Swiss would react to so blatant a violation of their neutrality. In any event, it would have been pointless: Heisenberg had visited Niels Bohr in Copenhagen in 1941 and showed him a crude device that Bohr took to be his design for a nuclear reactor. Two years later, when Bohr fled Denmark, he went to Los Alamos and recapitulated the design for American nuclear scientists working on the bomb; they immediately realized the device wouldn't work. The Germans had taken a wrong turn and had no hope of building a bomb.

Heisenberg and the German atomic bomb project: Jeremy Bernstein, ed., *Hitler's Uranium Club* (Woodbury, N.Y., 1996).

Heisenberg's "uncertainty principle": Samuel Goudsmit, *Alsos* (New York, 1947).

Heisenberg 1945, Farm Hall: David Cassidy, *Uncertainty* (Basingstroke, Eng., 1992).

Leo Szilard: Leo Szilard, "Reminiscences," *Perspectives in American History* II (1968).

Szilard and nuclear weapons: Bernard T. Feld, "Einstein and Nuclear Weapons," in Gerald Holton and Yehuda Elkana, eds., *Albert Einstein* (Princeton, N.J., 1982).

Einstein letter to President Roosevelt: Ronald W. Clark, *Einstein* (New York, 1971).

HURRY UP: Jeremy Bernstein, "Creators of the Bomb," *New York Review of Books* (May 11, 2000).

Manhattan Project: Richard Rhodes, *The Making of the Atomic Bomb* (New York, 1986). It should be noted that only after plans for Los Alamos had been completed did someone notice an interesting omission: no plans for any chapels or houses of worship.

J. Robert Oppenheimer: Murray Kempton, "The Ambivalence of J. Robert Oppenheimer," *Esquire* (December 1983).

Oppenheimer and Groves: Nuel Pharr Davis, *Lawrence and Oppenheimer* (New York, 1968).

1944 crisis: Federation of American Scientists, *FAS Public Interest Report*, "Conscience, Arrogation, and the Atomic Scientist" (July–August 1994). At least three scientists and one technician took the most radical step: espionage. Deciding that the bomb should be shared with the Soviet Union, they conveyed the secrets of the bomb's design and technical details to Soviet intelligence. Their actions saved the Soviets nearly ten years of required research and development. The first Soviet atomic bomb was a virtual copy of the "fat man" American bomb that was used on Hiroshima.

Trinity, Hiroshima: Gerhard L. Weinberg, *A World at Arms* (New York, 1992). Only two technologies useful for civilian life arose from the Manhattan Project, and one of them came about quite by accident. The first was the atomic reactor to generate power, although that remains problematic because of safety concerns and disposal of spent nuclear fuel. The second was a new miracle chemical substance, which actually came into being accidentally in 1938 when a DuPont chemist, experimenting with gaseous tetrafluorethylene in an attempt to develop a nontoxic refrigerant, discovered that when it hardened, the result was a strange solid material that was more inert than sand, yet unaffected by heat or solvents. It turned out to be perfect for gaskets to resist the corrosive uranium hexafluoride, one of the materials used in the production of U-235. DuPont realized it was on to something, and produced a commercial version for a variety of uses, including pacemakers, since the substance is one of the few that the human body's immune system does not reject. DuPont called this version of the substance Teflon. See "The Original Teflon Man," *The Chemist* (May 1985).

Oppenheimer post–World War II: J. H. Manley, "In the Matter of the H-Bomb," *Bulletin of the Atomic Scientists* (January 1984).

Stalin and Soviet science: Mikhail Heller and Aleksandr M. Nekrich, *Utopia in Power* (New York, 1986).

Soviet 1929 Five-Year Plan: W. W. Leontieff, "Scientific and Technological Research in Russia," *American Slavic Review* 4 (1945).

Stalin purge of the USSR Academy of Sciences: Andrei Sakharov, *Memoirs* (New York, 1990).

Stalin interference with science, Lysenko: Valery N. Soyfer, *Lysenko and the Tragedy of Soviet Science* (New Brunswick, N.J., 1994).

Soviet agricultural disaster: Zhores Medvedev, *The Rise and Fall of T.D. Lysenko* (New York, 1969).

Peter Kapitsa and the Soviet atomic bomb: Lawrence Badash, *Kapitsa, Rutherford, and the Kremlin* (Santa Barbara, Calif., 1985).

Condon Loyalty Board hearing: Carl Sagan, *The Demon-Haunted World* (New York, 1996).

Oppenheimer 1951 lecture: Kempton, op. cit.

9: The Age of Doom

Lang Chi, American "wonder weapons": Rand Corporation, Rand Report 1312-1, "The Laser-Guided Bomb" (June 1974).

Bush proposal: Dr. Vannevar Bush, director, OSRD, "Report to the President: Science, the Endless Frontier" (July 1945).

National Science Foundation: Meg Greenfield, "Science Goes to Washington," *The Reporter* (September 26, 1963).

National Defense Education Act: Roger Geiger, "Sputnik and the Academic Revolution," paper presented at Conference on Federal Support for University Research, University of California at Berkeley (May 1998).

Current federal funding of science: Daniel Sarewitz, *Frontiers of Illusion: Science, Technology, and Politics* (Philadelphia, 1996).

Fruits of federal funding of science: D. Allan Bromley, "Science and Surpluses," *New York Times* (March 9, 2001).

***Scientific American, Progressive* cases:** Mary M. Cheh, "The *Progressive* Case and the Atomic Energy Act: Waking to the Dangers of Government Information Controls," *George Washington University Law Review* (January 1980).

First hydrogen bomb test 1952: T. A. Heppenheimer, "What Edward Teller Did," *Invention and Technology* (Winter 2001).

"Thank you, Dr. Teller . . .": Quoted in Carl Sagan, *The Demon-Haunted World* (New York, 1996).

Increasing concern over effects of science: Brian L. Silver, *The Ascent of Science* (Oxford, 1998).

Andrei Sakharov: Andrei Sakharov and Edward Lazansky, *Andrei Sakharov and Peace* (New York, 1985).

Lasers: "Just Thanck Planck," *The Economist* (December 7, 2000).

Transistors: William Shockley, *Electronics and Holes in Semiconductors* (New York, 1950).

Consequences of lasers and transistors: International Institute for Strategic Studies, *Adelphi Papers #118: Precision-Guided Weapons* (London, 1975).

Prime numbers, composites: C. H. Hardy, *A Mathematician's Apology* (Cambridge, 1940).

National Security Agency and computers: "Progress Made on Unbreakable Code," *Science Digest* (April 28, 2000).

Military and computers: George and Meredith Friedman, *The Future of War: Power, Technology, and American World Dominance in the 21st Century* (New York, 1996).

Modern science and military power: Barry Commoner, *Science and Survival* (New York, 1967).

Failure of Soviet science: Robert Wright, "The Experiment That Failed," *National Review* (October 28, 1991).

Chinese scientific revival: Richard Baum, "Chinese Science after Mao," *Wilson Quarterly* (Spring 1983).

Saddam Hussein: Roy A. Griggs, "Technology and Strategy," *Airpower Journal* (Winter 1999).

New military technology: Joseph Miranda, "War and the Military Revolution in the 21st Century," *Strategy and Tactics* (2000).

Afterword: Of Microbes and Thunderbolts

Dachau experiments: International Military Tribunal, *Trials of War Criminals: The Medical Cases*, vol. 1 (October 1946–April 1949).

Nuremberg conference: Benno Mueller-Hill, *Murderous Science* (New York, 1988).

Mating of science and the state: Dudley R. Hershbach, "Imaginary Gardens with Real Toads," *Annals of the New York Academy of Sciences* 775 (1996).

Japanese advances in epidemiology: Louis Livingston Seaman, *The Real Triumph of Japan: The Conquest of the Silent Foe* (Philadelphia, 1906).

Shiro, Unit 731: Sheldon H. Harris, "Japanese Biological Warfare Research on Humans: A Case Study of Microbiology and Ethics," *Annals of the New York Academy of Sciences* 666 (1992).

U.S. failure to exempt science students: U.S. Joint Chiefs of Staff report, "Statistical Report of Aliens Brought to the United States under the Paperclip Program" (December 1, 1952).

American shock over German, Japanese science: Frank Malina, "The Jet Propulsion Laboratory," in Eurgeme M. Emme, ed., *History of Rocket Technology* (Bloomington, Ind., 1964); P. E. Cleator, "Autopsia," *Journal of the British Interplanetary Society* (May 1948).

Operation Paperclip: Tom Bower, *The Paperclip Conspiracy* (London, 1987).
Strughold and Paperclip: Charles R. Allen Jr., "Hubertus Strughold, Nazi in the USA: Atrocities in the Name of Medical Science," *Jewish Currents* (December 1974).
Strughold "father of space medicine": National Aeronautics and Space Administration Report SP-4201, "Beginnings of Space Medicine" (1981).
U.S. immigration list, cover-up: "U.S. Immigration Agency Lists 37 in Inquiry on War Crimes," *New York Times* (June 6, 1974); Linda Hunt, "U.S. Cover-up of Nazi Scientists," *Bulletin of the Atomic Scientists* (April 1985). Strughold died in 1986. Subsequently, as information about his wartime record became public, his honors were stripped from his memory. In 1993 his name was removed from a mural of "medical heroes" at Ohio State University, and two years later the U.S. Air Force removed his name from the School of Aerospace Medicine, which originally had been named in his honor.
Dr. Arthur Rudolph: Robert Sherrill, "The Golden Years of an Ex-Nazi," *Nation* (June 7, 1986). In 1984 the U.S. Justice Department sought to revoke Rudolph's American citizenship on the grounds he was a Nazi war criminal and thus had obtained that citizenship illegally. Although Nazi war crimes are not violations of American law, the Displaced Persons Act of 1947, under which the Paperclip assets were admitted to the United States and granted citizenship, specifically barred members of the Nazi Party, SS, and the Gestapo, along with any persons involved in "crimes against humanity." According to the law, anyone found to have obtained U.S. citizenship under the act who has concealed membership in Nazi organizations faces revocation of citizenship and deportation. Rudolph voluntarily relinquished his citizenship and returned to Germany. The government's move against Rudolph can be interpreted cynically: By 1984, the "rocket team" had long outlived its usefulness and had been squeezed out during a drive to "Americanize" the U.S. military rocket and space programs.
Dora concentration camp: Yad Vashem Document TR 10/769, *Landgericht Essen* (Jerusalem, 1972).
"We do not at this time . . .": International Military Tribunal for the Far East in Tokyo, "Trial of the Major Japanese War Crimes" (August 29, 1946).
Russian trial: Military Tribunal of the Primorye Military Area, "Materials on the Trial of Former Servicemen of the Japanese Army Charged with Manufacturing and Employing Bacteriological Weapons" (December 25–30, 1949).
Dr. Walter Schreiber: Paris Jewish Contemporary Documentation Center, *Les Médecins du Mort* [*The Doctors of Death*] (Paris, 1976).

Russian crash CBW program: Ken Alibek, *Biohazard* (London, 1999). The American program also had its share of accidents, including a malfunction at a Utah testing area in 1968, when an accidental release of a toxic substance during an open-air test killed several thousand sheep.

Soviet rocket program: National Aeronautics and Space Administration, Report SP-2000-4408, "Challenge to Apollo: The Soviet Union and the Space Race, 1945–1974" (Washington, D.C., 2000).

Sergei Korolev: Alan Wells, "Father of Soviet Space," *New Scientist* (July 19, 1997).

American experiments on humans: Tim Beardsley, "The Cold War's Dirty Secrets," *Scientific American* (May 1995); "Without Consent," *New Scientist* (November 13, 1999); Jonathan Moreno, "Lessons Learned: A Half-Century of Experimenting on Humans," *The Humanist* (September–October 1999).

National Security Agency: Private information. Among the promising mathematical students the NSA sought to recruit was Theodore Kaczynski, who rejected the offer. The incident began a process in Kaczynski's mind that convinced him the government, in league with science and technology, had embarked on a plan to control the minds of all citizens. It would result, ultimately, in his withdrawal from society and a series of bombs he set off against this perceived megalith.

Indian nuclear weapons: "Deadly Secrets," *Scientific American* (August 1998).

Human Genome Project: Mikulas Teich, "The 20th Century Scientific-Technical Revolution," *History Today* (November 1996).

Genetic weapons: Charles Piller and Keith R. Yamamoto, *Gene Wars* (New York, 1988).

Robotics, nanotechnology: Bill McKibben, "Playing Pandora," *Washington Monthly* (June 2000).

Scope of modern science: Gerald Holton, *The Advancement of Science and Its Burden* (Cambridge, Mass., 1986).

Benjamin Franklin in Paris: I. Bernard Cohen, *Science and the Founding Fathers* (New York, 1995).

Index

Abu Bakr, 55
Adrianople, battle of, 36, 39
Africa, slave trade, 97
Agincourt, battle of, 38–41, 46, 61, 66, 84, 91
agriculture
  fertilizer, 153–154
  Soviet Union, 199–200
airplane
  altitude problems, 222–223, 231–232
  fighter plane, 175
  invention of, 165–166
  jet aircraft, 172, 230
  nosedives, 161
  science and, 166–167
  World War I, 166
airship
  Graf Zeppelin II, 175
  World War I, 164–165
air-to-ground missiles, 207
Albert, Charles d', 40
alchemy, gunpowder, 76–77
Alesia, siege of, 34
Alexander the Great, 29–30, 31–32, 33
Alexander VI (pope of Rome), 106–107
Alexandria (Egypt), 30–31, 47
Alfonso VII (king of Spain), 56
altitude problems (airplanes), 222–223, 231–232
anthrax, 235
antiaircraft missiles, 230
anti-Semitism
  Germany, 185–187, 224
  Soviet Union, 201
Apollo spacecraft, 233
Appert, Nicholas, 133, 138

applied science, pure science versus, 9, 31–32, 78, 79, 80, 163, 167–168, 183–184, 191, 211
Aquinas, Saint Thomas, 57–58, 60
archery, longbow, 46, *see also* bow and arrow; longbow
Archimedes of Syracuse, 32–33, 57, 80, 114, 148
Aristarchos, 31
Aristotle, 23, 24, 29, 31–32, 57, 78, 79, 95
Armenia, iron, 18
armor, *see also* knights
  Agincourt, battle of, 39–41
  Athens, 24–25
  costs of, 44
  crossbow, 51
  European, 89
  horse armor, 51
  Incas, 88
  technology of, 49–50
arms race, *see also* warfare
  artillery, 150–151
  industrialization, 147
  machine gun, 151
  naval, 147–149
  science and, 75
  World War I, 152–153
army, *see also* foot soldier
  France, 118–119, 124–125
  Prussia, 125
  standing armies, 75, 118
arquebus (musket), 83, 89, 113
artificial atmospheres, 232
artillery, *see also* cannon; specific artillery weapons
  arms race, 150–151

267